Lecture Notes in Mathematics

Edited by A. Dold ar

549

Brauer Groups

Proceedings of the Conference Held at Evanston, October 11–15, 1975

Edited by D. Zelinsky

Springer-Verlag
Berlin · Heidelberg · New York 1976

Editor
Daniel Zelinsky
Northwestern University
Department of Mathematics
Evanston, Il. 60091/USA

Library of Congress Cataloging in Publication Data

Main entry under title:

Brauer groups.

(Lecture notes in mathematics ; 549)
"Sponsored by Northwestern University."
Bibliography: p.
Includes index.
1. Brauer group--Congresses. 2. Separable algebras--Congresses. I. Zelinsky, Daniel. II. Northwestern University, Evanston, Ill. III. Series: Lecture notes in mathematics (Berlin) ; 549.
QA3.L28 no. 549 [QA251.3] 510'.8s [512'.24]
76-46416

AMS Subject Classifications (1970): 13A20, 16A16, 18H20, 14C20, 14H99, 14L15, 18D10

ISBN 3-540-07989-0 Springer-Verlag Berlin · Heidelberg · New York
ISBN 0-387-07989-0 Springer-Verlag New York · Heidelberg · Berlin

Printed in Germany
Printing and binding: Beltz Offsetdruck, Hemsbach/Bergstr.

CONTENTS

The Conference on Brauer Groups was originally titled Conference on Brauer and Picard Groups. The present title is more nearly representative of the contents of the conference and these proceedings.

The conference was sponsored by Northwestern University and was held there (Evanston, Illinois) from October 11 to 15, 1975.

The list of participants which follows gives the university of each participants at the time of the conference (Department of Mathematics in each case). Professor Chase had to cancel his attendance but kindly submitted his manuscript for these Proceedings.

Besides the papers published here, the following were read:

R. T. Hoobler	How to construct U. F. D.'s
M. Ojanguren	Generic splitting rings
S. Rosset	Some solvable group rings are domains
D. Haile	Generalization of involution for central simple algebras of order m in the Brauer group
S. A. Amitsur	Cyclic splitting of generic division algebras
G. Szeto	Lifting modules and algebras

CONFERENCE ON BRAUER GROUPS

List of Participants

AMITSUR, S. A.
Hebrew University
Jerusalem, Israel

AUSLANDER, Bernice
Univ. of Massachusetts
Boston, Massachusetts

CHASE, Stephen U.
Cornell University
Ithaca, New York

CHILDS, Lindsay N.
SUNY at Albany
Albany, New York

COOK, P. M.
Michigan State Univ.
East Lansing, Michigan

DEMEYER, Frank R.
Colorado State Univ.
Ft. Collins, Colorado

ELGETHUN, Edward
Univ. of N. Florida
Jacksonville, Florida

GARFINKEL, Gerald S.
New Mexico St. Univ.
Les Cruces, New Mexico

GUSTAFSON, William
Indiana University
Bloomington, Indiana

HAILE, Darrell
Northwestern Univ.
Evanston, Illinois

HOOBLER, Raymond
City College, CUNY
New York, New York

INGRAHAM, Edward
Michigan State Univ.
East Lansing, Michigan

KNUS, Max-Albert
ETH
Zurich, Switzerland

KREIMER, H. F.
Florida State Univ.
Tallahassee, Florida

LAFOLLETTE, Philip
Indiana Univ.
Bloomington, Indiana

LEE, Hei-Sook
Queens University
Kingston, Ontario

MAGID, Andy R.
Univ. of Illinois
Urbana, Illinois

OJANGUREN, Manuel
Westfälische-Wilhelms
University
Münster, Germany

ORZECH, Morris
Queens University
Kingston, Ontario

PAREIGIS, Bodo
Ludwig-Maximilians
University
Munchen, Germany

REINER, Irving
Univ. of Illinois
Urbana, Illinois

ROSSET, Shmuel
Tel Aviv University
Tel Aviv, Israel

SALTMAN, David J.
Yale University
New Haven, Conn.

SMALL, Charles
Queens University
Kingston, Ontario

SZETO, George
Bradley University
Peoria, Illinois

ZELINSKY, Daniel
Northwestern Univ.
Evanston, Illinois

ON BRAUER GROUPS OF SOME NORMAL LOCAL RINGS

Lindsay N. Childs

Let R be a Noetherian local domain with quotient field K, and let Br(R) denote the Brauer group of equivalence classes of Azumaya R-algebras. We are interested in when the map from Br(R) to Br(K) is one-to-one. For an earlier survey of this question, see [7].

The classic result on this question is M. Auslander and O. Goldman's result [4] that Br(R) → Br(K) is 1-1 if R is regular. B. Auslander [3] generalized this result, for R normal, to yield a description of the kernel of Br(R) → Br(K) as a factor group of the monoid of reflexive R-modules whose endomorphism rings are finitely generated projective R-modules. Computation of the kernel using B. Auslander's description is difficult, however. In [8] G. Garfinkel, M. Orzech and the author gave a simplified exposition of the above two results, and applied B. Auslander's description to obtain a more manageable criterion, Theorem 1.1 below, for showing that Br(R) → Br(K) is 1-1. Using it, we obtained examples of normal, local, non-regular R for which Br(R) → Br(K) is still 1-1.

The present paper summarizes some of [8], giving different proofs where possible, and extends [8] to obtain additional examples of rings R for which Br(R) → Br(K) is 1-1. In

---

This work was partially supported by the N.S.F.

particular, we note a class of examples for which the kernel of the map from the cohomological Brauer group of R to Br(K) is non-trivial and torsion-free.

1. The criterion

For R a normal Noetherian domain (or more generally a Krull domain), Cl(R), the divisor class group of R, is the free abelian group on the minimal prime ideals of R modulo the subgroup of principal ideals. Cl(R) measures the failure of factoriality of R [5]. For R normal and local, the Henselization R^h and strict Henselization R^{sh} of R are both normal and faithfully flat over R. Thus $Cl(R^h)$ and $Cl(R^{sh})$ are defined and the maps into them from Cl(R) are 1-1. We identify Cl(R) with its image in $Cl(R^h)$, resp. $Cl(R^{sh})$, below.

__Theorem 1.1__ Let R be a normal Noetherian local domain with maximal ideal m, quotient field K, and strict Henselization R^{sh}. Then there is a one-to-one homomorphism from $\ker\{Br(R) \to Br(K)\}$ to $Cl(R^{sh})/Cl(R)$.

In [8] this result was proved by applying the snake lemma to two applications of the exact sequence of B. Auslander. The proof here will follow Grothendieck [14].

__Proof.__ Let X = Spec(R), x = Spec(K), i: x → X, and define the sheaves (in the etale topology over X) $\underline{G}_{m,X}$, the multiplicative group, $\underline{R}^*_X = i_*\underline{G}_{m,x}$, the sheaf of invertible rational functions, and $\underline{Div}_X = \underline{R}^*_X/\underline{G}_{m,X}$, the sheaf of Cartier divisors. Then [14, Prop. 1.7] if H^*_{et} denotes etale cohomology, one has

$$\ker\{H^2_{et}(X,\underline{G}_{m,X}) \to H^2_{et}(X,\underline{R}^*_X)\} \cong H^1_{et}(X,\underline{Div}_X)\ .$$

Denote $H^2_{et}(X,\underline{G}_m) = Br'(X) = Br'(R)$, the cohomological Brauer group.

Claim: $\ker\{Br'(R) \to Br(K)\} = \ker\{Br'(R) \to H^2_{et}(X,\underline{R}^*_X)\}$.

For the Leray spectral sequence

$$H^p(X, R^q i_*(\underline{R}^*_X)) \Rightarrow H^n(x, \underline{R}^*_X)$$

yields a five term exact sequence of low degree

$$0 \to H^1(x, R^0 i_*(\underline{R}^*_X)) \to H^1(x, R^*_X) \to H^0(X, R^1 i_*(R^*_X)) \to$$
$$\to H^2(X, R^0 i_*(\underline{R}^*_X)) \to H^2(x, R^*_X) .$$

Now $R^1 i_*(R^*_X)) = 0$ (see [14], proof of Lemma 1.6),

$R^0 i_*(\underline{R}^*_X) = i_*\underline{R}^*_X = i_*\underline{G}_{m,x} = \underline{R}_X$, and $H^2(x, \underline{R}^*_X) = H^2(x,\underline{G}_{m,x}) = Br'(K) = Br(K)$

since $\underline{R}^*_X = \underline{G}_{m,x}$ in the etale topology. Thus $H^2(X, \underline{R}^*_X)$ maps 1-1 to $Br(K)$ and $\ker\{Br'(R) \to Br(K)\} = \ker\{Br'(R) \to H^2(X,\underline{R}^*_X)\}$. Finally, for $X = Spec(R)$, R local,

$$H^1(X, \underline{\underline{Div}}_X) \cong Cl(R^{sh})^G/Cl(R)$$

where $G = Gal(\bar{k}/k)$, $k = R/m$, by [14], Remark 1.11b. Thus $\ker\{Br'(R) \to Br(K)\} \subseteq Cl(R^{sh})^G/Cl(R)$. Now $Br(R)$ maps 1-1 to $Br'(R)$ by [13], (see [18], chapter 13, for an exposition of this), so the theorem follows.

Note that if R/m is algebraically closed, $R^h = R^{sh}$ and we get $\ker\{Br(R) \to Br(K)\} \subseteq Cl(R^h)/Cl(R)$. If the completion $\hat{R}$

of R is normal, then we can replace R^h by $\hat{R}$ in Theorem 1.1 since $Cl(R^h)$ maps 1-1 to $Cl(\hat{R})$.

2. Applications

For rings R satisfying the following definition it is clear from Theorem 1.1 that $Br(R) \to Br(K)$ is 1-1.

Definition. R is geometrically factorial if R^{sh} is factorial, i.e. $Cl(R^{sh}) = 1$.

Examples.

2.1 If R is regular, $Br(R) \to Br(K)$ is 1.1.

For R^{sh} is regular, hence factorial by the well-known result of M. Auslander and D. Buchsbaum.

2.2 Let R be a two-dimensional normal local ring with algebraically closed residue field k of characteristic not 2,3,5 such that the completion of R has maximal ideal generated by x,y,z with $x^2 + y^3 + z^5 = 0$. Then $Br(R) \to Br(K)$ is 1-1.

For Lipman [17] proves that such rings are geometrically factorial.

Lipman also proves that the examples 2.2 and analogous ones in char. ≤ 5 are the only two-dimensional geometric examples with $k = \bar{k}$ which are geometrically factorial.

2.3 Hoobler's paper in these proceedings gives criteria for finding higher-dimensional examples of geometrically factorial rings.

3. Perhaps the simplest example of a ring R with $Cl(R^{sh}) \neq 1$ is $R = (\mathbb{C}[x,y,z]/(x^2 + y^2 + z^2))_{(x,y,z)}$. From Lipman [17] one

has that $Cl(\hat{R}) \cong Cl(R^h) \cong Cl(R) \cong Z_2$; (that $Cl(R) = Z_2$ also follows from Theorem 5.1 below, given Lipman's computation of $Cl(R^h)$). Since $R/m = \mathbb{C}$, $Br(R) \to Br(K)$ is 1-1.

The rest of this paper is devoted to generalizations of this example.

4. The only explicitly known examples where $Br(R) \to Br(K)$ is not 1-1 are typified by the example

$$R = (\mathbb{R}[x,y,z]/(x^2 + y^2 + z^2))_{(x,y,z)} .$$

Then the class of the usual quaternion algebra $A = (\frac{-1,-1}{R})$ maps to the trivial element of $Br(K)$ because in K, -1 is the sum of two squares; but $A \otimes_R R/m = A \otimes_R \mathbb{R}$ is the usual real quaternions, hence A is non-trivial in $Br(R)$. (For similar kinds of real phenomena see DeMeyer's article in these proceedings.) These examples are not "geometric" in that the field of constants is not algebraically closed, and, in particular, has a non-trivial Brauer group.

Note here that $Cl(R) = 1$, necessarily by Theorem 1.1. This is of course well-known (see [11], Prop. 11.5), (see also 2.5, 2.6, of [8]).

For the rest of the paper we assume that $R/m = \mathbb{C}$, so that $R^{sh} = R^h$.

5. Given that the completion of the ring

$$S = (\mathbb{C}[x,y,z]/(x^2 + y^2 + z^2))_{(x,y,z)}$$

has class group Z_2, natural candidates for rings R with $Br(R) \to Br(K)$ not 1-1 are local rings which are factorial but whose completions are isomorphic to $\hat{S}$. We consider therefore

$$R = (\mathbb{C}[x,y,z]/(x^2 - ux^2 - vy^2))_{(x,y,z)}$$

where u,v are in $\mathbb{C}[x,y]$ with non-zero constant terms, so that u,v are units of R. Then in $\hat{R}$,u and v have square roots s,t, so $\hat{R} \cong \mathbb{C}[[x,y]][z]/(x^2 - s^2x^2 - t^2y^2)$ and $(x,y) \to (sx,ty)$ is an invertible change of coordinates yielding an isomorphism of $\hat{R}$ and $\hat{S}$.

Theorem 5.1. The following conditions are equivalent.

a. R is not factorial

b. $Br(R) \to Br(K)$ is 1-1

c. The quaternion algebra $(\frac{u,\ v}{R})$ is trivial in Br(R)

d. $1 = u\alpha^2 + v\beta^2$ has a solution in R.

Proof. a => b follows from Theorem 1.1, given that $Cl(\hat{S}) = Z_2$.

c<=> d: The quaternion algebra $(\frac{u,\ v}{R}) = (\frac{\frac{1}{u},\ -v/u}{R})$ is trivial iff 1/u is a norm from $T = R[t]/(t^2 + v/u)$ (view $(\frac{\frac{1}{u},\ v/u}{R})$ as a cyclic crossed product $D(T, Z_2)_{1/u}$), iff $\frac{1}{u} = \beta^2 \frac{1}{v} + \alpha^2$ iff $1 = \alpha^2 u + \beta^2 v$.

b => c: By c<=>d and the fact that $1 = u(\frac{x}{z})^2 + v(\frac{y}{z})^2$, $(\frac{u,\ v}{K})$ must be trivial in Br(K). If Br(R) → Br(K) is 1-1 then $(\frac{u,\ v}{R})$ must be trivial in Br(R).

d => a: if $1 = u\alpha^2 + v\beta^2$, $x^2u + y^2v = z^2u\alpha^2 + z^2v\beta^2$, so $u(z\alpha + x)(z\alpha - x) = v(y + z\beta)(y - z\beta)$. Since both sides are in $m^2 - m^3$, this is an example of failure of factoriality in R. That completes the proof.

We have shown that for this class of examples, knowing ker{Br(R) → Br(K)} is equivalent to knowing Cl(R).

Here is a useful result relating to factoriality of R.

Lemma 5.2. With R as above, suppose there exist A,B,C, in the maximal ideal m_o of $\mathbb{C}[x,y]_{(x,y)}$ so that

$$ux^2 + vy^2 = B^2 + AC .$$

Then R is not factorial.

Proof. Since $z^2 = ux^2 + vy^2 = B^2 + AC$, we get a non-unique factorization $(z + B)(z - B) = AC$. For A,B,C must all be in $m_o - m_o^2$, hence all factors are irreducible.

We note that a necessary and sufficient condition that R be not factorial is that $z^2 - ux^2 - vy^2 = \det D$, where D is an n×n matrix, $n \geq 2$ (in our case, n = 2 necessarily), with entries in the maximal ideal of $(\mathbb{C}[x,y,z])_{(x,y,z)}$. This is a special case of a result of Andreotti and Salmon [1], see Eisenbud [11]. If the hypothesis of lemma 5.2 holds, then $z^2 - ux^2 - vy^2 = \det \begin{pmatrix} z-B & C \\ A & z+B \end{pmatrix}$, so we obtain the Andreotti-Salmon condition easily. I don't know if the converse to lemma 5.2 holds.

Using lemma 5.2 we get

Theorem 5.3. Let $R = (\mathbb{C}[x,y,z]/(z^2 - ux^2 - vy^2))_{(x,y,z)}$ and suppose that u,v are polynomials in $\mathbb{C}[x,y]$ with non-zero constant terms and of degree ≤ 2. Then $Br(R) \to Br(K)$ is 1-1.

Proof. We find A,B,C in $\mathbb{C}[x,y]$ with

$$(1) \qquad W = x^2u + y^2v = B^2 + AC .$$

Let $C = x - ty$. Then there exists A,B with (1) holding iff there exists $B(x,y)$ with $W(ty,y) = B(ty,y)^2$. Write $W(x,y) = W_2(x,y) + W_3(x,y) + W_4(x,y)$, where $W_i(x,y)$ is the sum of the terms in $W(x,y)$ of total degree i, and similarly for B: $B = B_1 + B_2$. Then (1) holds if there exists B with

$$(2) \qquad \begin{aligned} W_2(ty,y) &= B_1(ty,y)^2 \\ W_3(ty,y) &= 2B_1(ty,y)B_2(ty,y) \\ W_4(ty,y) &= B_2(ty,y)^2. \end{aligned}$$

Since the equations in (2) are homogeneous in y, they have solutions for t iff there is a t solving

$$W_2(t,1) = B_1(t,1)^2$$

(3) $$W_3(t,1) = 2B_1(t,1)B_2(t,1)$$

$$W_4(t,1) = B_2(t,1)^2.$$

Consider the equation

(4) $$4W_2(t,1)W_4(t,1) = W_3(t,1)^2 .$$

Since $W_2(t,1) = a + bt$ with $a = v(0,0)$, $b = u(0,0)$, $a \neq 0, b \neq 0$, equation (4) has a solution t in $\mathbb{C}$ unless $W_3(x,y) = cy^3$, $c \neq 0$ in $\mathbb{C}$ and $W_4(x,y)$ is identically zero. If this last situation holds, then $u = b$ in $\mathbb{C}$, and $ux^2 + vy^2 = (xb^{\frac{1}{2}})^2 + y(vy)$, satisfying lemma 5.2. Otherwise, let t in $\mathbb{C}$ be a solution of (4). Choose coefficients of $B_1(x,y) = b_{11}x + b_{12}y$ so that $W_2(t,1) = B_1(t,1)^2$ in $\mathbb{C}$. If $W_2(t,1) = 0$, choose coefficients of $B_2(x,y)$ so that $B_2(t,1)^2 = W_4(t,1)$. By (4), $W_3(t,1) = 0$ so (3) will be solved. If $W_2(t,1) \neq 0$, choose coefficients of $B_2(x,y)$ so that $2B_2(t,1) = W_3(t,1)B_1(t,1)^{-1}$, then $B_2(t,1)^2 = W_4(t,1)$ by (4), and (3) will be solved. This completes the proof of the theorem.

This result extends and corrects Theorem 7 of [8]. The latter claimed that $Br(R) \to Br(K)$ was 1-1 if u,v had degree ≤ 1, but the proof provided was valid only for a dense subset of u,v in $\mathbb{C}^6$.

Question: what about $z^2 = x^2 + x^3 + y^2 + y^5$? I am unable to show that if the defining relation for R is that equation, then R is not factorial. Indeed, this bears in a very naive sense a resemblance to Lipman's example!

6. Another way to view the example $R = (\mathbb{C}[x,y,z]/(x^2+y^2+z^2))_{(x,y,z)}$ is as the local ring at the vertex of the cone on the projective non-singular plane curve $F = x^2 + y^2 + z^2 = 0$. One can ask if the map from Br(R) to Br(K) is 1-1 for more general F. The affirmative answer is an immediate consequence of work of Danilov [9]. My thanks to S. Rosset for suggesting I look at Danilov's work.

<u>Theorem 6.1.</u> Let X be a smooth projective plane curve, let $F(x,y,z)$ be its homogeneous equation, let A be the affine ring of the cone on X: $A = \mathbb{C}[x,y,z]/(F)$. Let A_m be the local ring of the vertex $m = (x,y)A$. Let $\hat{A}_m$ be the completion. Then $\hat{A}_m$ is normal and $\mathrm{Cl}(\hat{A}_m) = \mathrm{Cl}(A_m) \oplus V$, where V is a finite dimensional complex vector space. Thus $\mathrm{Cl}(\hat{A}_m)/\mathrm{Cl}(A_m)$ is torsion-free, and $\mathrm{Br}(A_m) \to \mathrm{Br}(K)$ is 1-1.

<u>Proof.</u> The normality of $\hat{A}_m$ is [9], Lemma 4. Since we are over a characteristic zero field, [9], Proposition 8, gives that $\mathrm{Cl}(\hat{A}_m) = \mathrm{Cl}(A) \oplus \bigoplus_{n\geq 1} H^1(X, \mathcal{O}_X(n))$. By [19], §78, Prop. 5, $H^1(X, \mathcal{O}_X(n)) \cong \mathrm{Hom}_{\mathbb{C}}(H^0(X, \mathcal{O}_X(N-n)), \mathbb{C})$, where $N = g-3$, g = the degree of F. Thus if $g \leq 3$, $\mathrm{Cl}(\hat{A}_m) \cong \mathrm{Cl}(A)$; if $g \geq 4$

$Cl(\hat{A}_m) \cong Cl(A) \oplus \bigoplus_{n=1}^{g-3} H^1(X, 0_X(n)) = Cl(A) \oplus$ a non-trivial finite dimensional complex vector space V.

Now $Cl(A) \cong Cl(A_m)$. For the 1-1 map from Cl(A) to $Cl(\hat{A}_m)$ factors through $Cl(A_m)$, so $Cl(A) \to Cl(A_m)$ is 1-1; ontoness follows from the surjectivity of Cl under taking rings of quotients, see [12], Corollary 7.2. Thus $Cl(\hat{A}_m)/Cl(A_m) = V$ is torsion free. Since ker $\{Br(A_m) \to Br(K)\}$ is torsion, it is trivial.

Remark. For $R/m = \mathbb{C}$, Theorem 1.1 shows that $\ker\{Br'(R) \to Br(K)\} \cong Cl(R^h)/Cl(R)$, where Br' is the cohomological Brauer group. That $Cl(R^h) \cong Cl(\hat{R})$ in the situations described in Theorem 6.1 is known by Danilov [10], §5, p. 235.
Thus for plane curves F of degree ≥ 4, Theorem 6.1 provides examples of non-triviality of $\ker\{Br'(R) \to Br(K)\}$. The previous existence of an example where $Br'(R) \to Br(K)$ was not 1-1 was known by Grothendieck ([14], 1.11b) based on an example of Mumford. Grothendieck's subsequent comment: "il ne devrait pas etre difficile de la meme facon de construire un exemple d'une surface algebrique normal pour laquelle $Br(X) \to Br(K)$ n'est pas injectif" remains unexplicated.

7. For those acquainted with Lipman's work [17] we remark that $Cl(R^h)/Cl(R)$ arises there as follows. Let R be a two-dimensional normal local ring with $R/m \cong \mathbb{C}$. Then [17,(16.3)] there is a

commutative diagram with exact rows

$$
\begin{array}{ccccccccccc}
0 & \to & \mathrm{Pic}^\circ(R) & \to & \mathrm{Cl}(R) & \to & H(R) & \to & G(R) & \to & 0 \\
 & & \downarrow & & \downarrow & & \downarrow & & \downarrow & & \\
0 & \to & \mathrm{Pic}^\circ(R^h) & \to & \mathrm{Cl}(R^h) & \to & H(R^h) & \to & G(R^h) & \to & 0
\end{array} \tag{7.1}
$$

Here $H(R)$ is a finite group of order $\det((E_iE_j))$ where (E_iE_j) is the intersection matrix of the exceptional locus of a desingularization of $\mathrm{Spec}(R)$.

From [17,(14.4)], $G(R^h) = 0$. Form [17,§17] one has that if $\mathrm{Cl}(R)$ is finite (i.e. R has a rational singularity) then $\mathrm{Pic}^\circ(R) = 0$. Thus if $\mathrm{Cl}(R^h)$ is finite, $\mathrm{Pic}^\circ(R) = \mathrm{Pic}^\circ(R^h) = 0$, and (7.1) yields that $G(R) \cong \mathrm{Cl}(R^h)/\mathrm{Cl}(R)$. This isomorphism applies to the examples of Section 5 above.

8. One interesting aspect of all examples considered in this paper is that they are local rings of normal singularities of surfaces which can be resolved by a single blowup. Thus the examples considered here are of interest because they relate to the question:

What happens to Br under a blowup? Is it 1-1?

One way to understand $\ker\{\mathrm{Br}(R) \to \mathrm{Br}(K)\}$ for R the affine ring of a complex surface might be to resolve the singularities of the surface. Since the map from the Brauer group of a non-singular surface to the Brauer group of its function field is 1-1 (and more then that is known: see [2]), any kernel would appear during the resolution process. One desingularizes a surface by a finite sequence of blowups and normalizations. One has some information on what happens to Br under normalization, by means of a Mayer-Vietoris sequence [6], [16]. Thus an understanding of what happens to Br under a blowup is the major lacuna in understanding $\ker\{Br(R) \to Br(K)\}$ for R the affine ring of a singular surface.

REFERENCES

1. A. Andreotti, P. Salmon, Annelli con unica decomponibilita in fattori primi ed un problema di intersezioni complete, Monatsh. für Math. 61(1957), 97-142.

2. M. Artin, D. Mumford, Some elementary examples of unirational varieties which are not rational, Proc. London Math. Soc. 25 (1972), 75-95.

3. B. Auslander, The Brauer group of a ringed space, J. Algebra 4 (1966), 220-273.

4. M. Auslander, O. Goldman, The Brauer group of a commutative ring, Trans. Amer. Math. Soc. 97(1960), 367-409.

5. N. Bourbaki, Algebre Commutative VII, Paris, Hermann, 1965.

6. L. Childs, Mayer-Vietoris sequences and Brauer groups of non-normal domains, Trans. Amer. Math. Soc. 196 (1974), 51-67.

7. _________, Brauer groups of affine rings, Ring Theory, Proc. Oklahoma Conf., New York, Marcel Dekker, 1974, 83-94.

8. L. Childs, G. Garfinkel, M. Orzech, On the Brauer group and factoriality of normal domains, J. Pure and Appl. Algebra 6 (1975), 111-123.

9. V.I. Danilov, The group of ideal classes of a completed ring, Math. USSR-Sbornik 6 (1968), 493-500.

10. _________, On rings with a discrete divisor class group, Math. USSR - Sbornik 17 (1972), 228-236.

11. D. Eisenbud, Some directions of recent progress in commutative algebra, Proc. Symp. Pure Math. 29 (1975), 111-128.

12. R. Fossum, The Divisor Class Group of a Krull Domain, Springer-Verlag, 1973.

13. A. Grothendieck, Le groupe de Brauer I, in Dix Exposés sur la cohomologie des schemas, North-Holland, 1968.

14. __________, Le groupe de Brauer II, loc. cit.

15. __________, Le groupe de Brauer III, loc. cit.

16. M. Knus, M. Ojanguren, A Mayer-Vietoris sequence for the Brauer group, J. Pure Appl. Algebra 5 (1974), 345-360.

17. J. Lipman, Rational singularities with applications to algebraic surfaces and unique factorization, Publ. Math. IHES 36 (1969), 195-279.

18. M. Orzech, C. Small, The Brauer Group of Commutative Rings, New York, Marcel Dekker, 1975.

19. J.-P. Serre, Faisceaux algebriques coherents, Ann. Math. 61 (1955), 197-278.

THE BRAUER GROUP OF AFFINE CURVES

by F. R. DeMeyer

Let k be a perfect field and $k[x_1, \ldots, x_n]$ the ring of polynomials in n-variables over k. Let I be an ideal of co-height = 1 in $k[x_1, \ldots, x_n]$ and let $R = k[x_1, \ldots, x_n]/I$. Then R is the coordinate ring of the affine curve X = Spec (R) over k. Let Br(R) be the Brauer group of Azumaya-algebras over R. A procedure is given for reducing the calculation of Br(R) to the calculation of Br(F) where F is the function field of an irreducible, non-singular curve. Then this procedure is applied to curves over the real numbers $\mathbb{R}$ and over finite fields. A complete exposition of the calculation of the Brauer group of real curves was given by M. A. Knus and the author in [3] which also contains the result for finite fields and some other remarks on the Brauer group and Picard group. We turn our attention to curves over arbitrary perfect fields. We show $Br(k[x, x^{-1}]) = Br(k) \oplus \hat{G}$ where $\hat{G}$ is the dual of the Galois group of the algebraic closure $\bar{k}$ of k, and some simple examples are calculated. Most of this paper is a consequence of joint work of the author and M. A. Knus.

Section I. As above let k be a perfect field and $R=k[x_1, \ldots, x_n]/I$ be the coordinate ring of the affine curve X over k. If we think of X as the points in $\bar{k}^{(n)}$ which are zeros of all the polynomials in I then the points $x \in X$ correspond to the maximal ideals M_x in R where M_x is the ideal of polynomial functions in R vanishing at x. Denote R/M_x by R(x). If A is an Azumaya R-algebra, let A(x) be the central simple R(x) algebra $R(x) \otimes A$. Sometimes one can think of an Azumaya R-algebra A as a parameterized system of central simple R(x) algebras A(x) as $x \in X$. We now give the reduction procedure for the calculation of Br(R).

Reduction to irreducible affine curves.

If N is a nilpotent ideal of R then Br(R) = Br(R/N) (Theorem 1 of [2]) and X = Spec R = Spec R/N so we can assume R has no nilpotent ideals. Thus (0) is the irredundant intersection of finitely many prime ideals, $(0) = \bigcap_{i=1}^{n} p_i$. If $S = \oplus \sum R/p_i$ then R can be viewed as a subring of S in a natural way. Let $c = \{x \in S \mid sx \in R \text{ for all } s \in S\}$ be the conductor from S to R. Then $c = \oplus \sum c_i$ where $c_i = \mathrm{Ann}_R p_i = \bigcap_{j \neq i} p_j$. We obtain the cartesian square

$$\begin{array}{ccc} R & \longrightarrow & S \\ \downarrow & & \downarrow \\ R/c & \longrightarrow & S/c \end{array}$$

which gives rise to a Mayer-Vietoris sequence ([1] or [5]) whose last four terms are

$$\text{(1.1)} \qquad \mathrm{Pic}(S/c) \to \mathrm{Br}(R) \overset{\phi}{\to} \mathrm{Br}(S) \oplus \mathrm{Br}(R/c) \overset{\psi}{\to} \mathrm{Br}(S/c).$$

The homomorphism ϕ associates to a class [A] in Br(R) the element $([S \otimes A], [R/c \otimes A])$ in $B(S) \oplus \mathrm{Br}(R/c)$. The homomorphism ψ associates to an element ([B], [C]) in $B(S) \oplus \mathrm{Br}(R/c)$ the element $[S/c \otimes_S B]$ times $[S/c \otimes_{R/c} C]^{-1}$ in Br(S/c). The terms in (1.1) have the following geometric interpretation: the rings R/p_i are the affine rings of the irreducible components X_i of X, the ideals c_i contain all the elements of R which vanish on the components X_j for $j \neq i$, in R/p_i the ideal c_i is the intersection of maximal ideals corresponding to points on X_i which also lie on some X_j for $j \neq i$. Thus c is the intersection of maximal ideals of R corresponding to points X which lie on two or more irreducible components of X. There are finitely many such points so S/c is semilocal and Pic(S/c) = 0. The Brauer group of S is the direct sum of the Brauer groups of the irreducible components X_i of X.

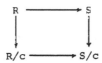

If x_ℓ is a point on two or more components of X the $R(x_\ell)$ is a finite field extension of k and $Br(R/c) = \oplus\sum Br(x_\ell)$. The maximal ideals of S containing c correspond to the points on each X_i which also lie in X_j for some $j \neq i$. Thus $Br(S/c) = \oplus \sum Br(R(x_\ell)^{n_\ell}$ where n_ℓ is the number of components X_i of X containing the point x. We want to describe ker ψ in (1.1). Let ψ_ℓ be ψ followed by the projection of $\oplus\sum Br(R(x_\ell))^{n_\ell}$ onto $Br(R(x_\ell))^n$. Then $\ker \psi = \bigcap \ker \psi_\ell$. Let $[B] = ([B_1], \ldots, [B_n]) \in Br(S)$ and $[C] \in Br(R/c)$, then $([B], [C]) \in \ker \psi_\ell$ if and only if for each j such that X_j contains x_ℓ we have $B_j(x_\ell) \otimes C(x_\ell)^0$ represents the zero class in $Br(R(x_\ell))$. (Here $C(x_\ell) = R(x_\ell) \otimes C$). We have proved the following.

(1.2) Let R be the coordinate ring of the affine curve X over the field k. Let $X_1, \ldots, X_n$ be the irreducible components of X and let R_i be the coordinate ring of X_i. Let A_i be an Azumaya R_i algebra $(i = 1, \ldots, n)$. Then $(A_1, \ldots, A_n)$ represents an element in Br(R) if and only if for any point $x \in X_i \cap X_j$ for $i \neq j$ we have $A_i(x) = A_j(x)$. Moreover, $(A_1, \ldots, A_n)$ represents the trivial class in Br(R) if and only if each A_i represents the trivial class in $Br(R_i)$.

Reduction to non-singular irreducible affine curves.

Let R be the coordinate ring of an irreducible curve X and let $\bar{R}$ be the integral closure of R. Then R is the coordinate ring of an affine desingularization Y of X. Let $c = \{x \in \bar{R} \mid \bar{R}x \subseteq R\}$ be the conductor from $\bar{R}$ to R. The maximal ideals in R containing c correspond to the singular points on X (we denote these points x_i) and the maximal ideals in $\bar{R}$ containing c correspond to the points on Y lying over the x_i (we denote these points y_{ij}). The cartesian square

$$\begin{array}{ccc} R & \longrightarrow & \bar{R} \\ \downarrow & & \downarrow \\ R/c & \longrightarrow & \bar{R}/c \end{array}$$

induces [1] or [5] the sequence

$$0 \to Br(R) \overset{\phi}{\to} Br(\bar{R}) \oplus Br(R/c) \overset{\psi}{\to} Br(\bar{R}/c)$$

which is exact since Pic $(\bar{R}/c) = 0$. As in section 1 the maps ϕ and ψ are the diagonal map and the difference map, respectively.

The Brauer group is unchanged by factoring by a nilpotent ideal (Theorem 1 of [2]) so we see that $Br(R/c) = \oplus \sum Br(R(x_i))$ and $Br(\bar{R}/c) = \oplus \sum Br(\bar{R}(y_{ij}))$. Thus we can describe the kernal of ψ.

(1.3) If $[B] \in Br(\bar{R})$ and $[C] \in Br(R/c)$ then $([B], [C])$ represents an element in $Br(R)$ if and only if $B(y_{ij}) \otimes (C(x_i) \otimes \bar{R}(y_{ij}))^0$ is in the zero class of $\bar{R}(y_{ij})$ for all points y_{ij} on Y lying over a singular point $x_i \in X$. $([B], [C])$ represents the trivial class in $Br(R)$ if and only if $[B]$ is trivial in $B(\bar{R})$ and $[C]$ is trivial in $Br(R/c)$.

Reduction to the Brauer group of the function field of an irreducible non-singular affine curve.

Let R be the coordinate ring of a non-singular irreducible affine curve X over the perfect field k. If X is not absolutely irreducible then R contains an extension K of k in the algebraic closure of k. We can assume X is defined over K and X is absolutely irreducible over K. Let Y be a regular completion of X so that Y - X contains only finitely many points. Let F be the function field of Y. The first terms of the exact sequence (2.1) of Grothendieck in [4], III for X and Y are

(1.4) $$0 \to Br(Y) \to Br(F) \overset{\tau}{\to} \oplus_Y H^1(y, Q/Z) \to H^3(Y, G_m) \to H^3(k, G_m)$$

(1.5) $$0 \to Br(X) \to Br(F) \overset{\tau}{\to} \oplus_X H^1(x, Q/Z) \to H^3(X, G_m) \to H^3(k, G_m)$$

where y (respectively x) runs through the closed points of Y (respectively X). We return to this pair of sequences when calculating the Brauer group of a curve over a finite field.

Section II. In [3] M. A. Knus and the author calculated the Brauer group of any affine curve over the real numbers $\mathbb{R}$. The outline sketched above together with the calculation of the Brauer group of the function field of real curves given by E. Witt in [7] was employed [3] to prove the following.

Theorem 2.1. Let I be an ideal of co-height one in $\mathbb{R}[x_1, \ldots, x_n]$, let $R = \mathbb{R}[x_1, \ldots, x_n]/I$, and let $X = \mathrm{Spec}(R)$ be the corresponding curve. If s is the number of real components of X in the strong topology then $\mathrm{Br}(R) = (Z/2Z)^s$.

The corresponding result for curves over finite fields is only indicated in [3] so we give a complete proof here.

Theorem 2.2. Let k be a finite field and let I be an ideal of co-height one in $k[x_1, \ldots, x_n]$. Let $R = k[x_1, \ldots, x_n]/I$, and let $X = \mathrm{Spec}(R)$ be the corresponding curve. Write $X = \bigcup_{i=1}^{n} X_i$ where X_i are the irreducible components of X. If n_i is one less than the number of closed points at infinity on a regular completion of the desingularization of X_i and $n = \sum_{i=1}^{n} n_i$ then $\mathrm{Br}(R) = (Q/Z)^n$.

Proof: We can assume R contains no nilpotent ideals. Let R_i be the coordinate ring of X_i, then by (1.1) an Azumaya algebra $(A_1, \ldots, A_n)$ over $R_1 \oplus \ldots \oplus R_n$ represents an element in Br(R) if and only if in $\mathrm{Br}(R(x))$, $[A_i(x)] = [A_j(x)]$ for each $x \in X_i \cap X_j$. Since k is finite, $\mathrm{Br}(R(x))$ is trivial so this condition is always satisfied and $\mathrm{Br}(R) = \mathrm{Br}(R_1) \oplus \ldots \oplus \mathrm{Br}(R_n)$. Let $\bar{R}_i$ be the integral closure of R_i, then $\bar{R}_i$ is the coordinate ring of a desingularization Y_i of X_i. For each singular point x on X_i and for each point y on Y_i lying over x the Brauer groups $\mathrm{Br}(R_i(x))$ and $\mathrm{Br}(\bar{R}_i(y))$ are trivial since $R_i(x)$ and $\bar{R}_i(y)$ are finite so by (1.2) we have $\mathrm{Br}(R_i) = \mathrm{Br}(\bar{R}_i)$. By replacing k by a finite extension if necessary we can assume Y_i is absolutely irreducible. Let $\overline{Y}_i$ be a regular completion of Y_i, then by [4], Remarque 2.5b. $Br(\overline{Y}_i) = 0$ and $H^3(\overline{Y}_i, G_m) = Q/Z$. By Hilbert's Theorem 90, $H^3(k, G_m) = 0$ and for any $y \in Y_i$, $H^1(y, Q/Z) = Q/Z$. The sequences (1.4) and (1.5) become

$$0 \to Br(F) \overset{\tau}{\to} \oplus_y (Q/Z)_y \to (Q/Z) \to 0$$

$$0 \to Br(Y_i) \to Br(F) \overset{\tau}{\to} \oplus_x (Q/Z)_x.$$

Comparing the two sequences we see the kernal of τ in the second sequence is $(Q/Z)^{n_i-1}$ where $n_i - 1$ is the number of (closed) points in $\overline{Y}_i - Y_i$. This proves the theorem.

Notice that there are Azumaya algebras A defined on an affine curve X over a finite field k with A(x) in the zero class of Br(R(x)) for all $x \in X$ yet A does not represent the zero class of Br(R). Several remarks on the behavior of the Brauer group and Picard group of real curves can be found in [3]. The Brauer group of a complex curve is easily seen to be trivial. By applying Tsen's theorem which asserts that the Brauer group of the function field of a curve over the complex numbers is trivial, one can show in the same way as in Theorem 2.2 that the Brauer group of the coordinate ring of any complex curve must be trivial (p. 89 of [4] III).

<u>Section III</u>. In this section we calculate some simple examples. The following result will be useful.

<u>Proposition 3.1</u>. Let k be a perfect field with Galois group G and let $\hat{G}$ be the dual of G. Let t be an indeterminate, then

$$B(k[t, t^{-1}]) = Br(k) \oplus \hat{G}.$$

<u>Proof</u>: By Tsen's theorem every element in $Br(k[t, t^{-1}])$ is split by $N[t, t^{-1}]$ where N is a normal field extension of k. If G(N/k) is the Galois group of N over k and $U(N[t, t^{-1}])$ is the group of multiplicative units in $N[t, t^{-1}]$ then

$$\begin{aligned}
Br(k[t, t^{-1}]) &= \varinjlim [H^2(G(N/k), U(N[t, t^{-1}]))] \\
&= \varinjlim [H^2(G(N/k), U(N) \times \langle t\rangle)] \\
&= \varinjlim [H^2(G(N/k), U(N)] \times H^2(G(N/k), \langle t\rangle)] \\
&= Br(k) \times H^2(G, Z) \text{ where } G \text{ acts trivially on } Z \\
&= Br(k) \oplus \hat{G}.
\end{aligned}$$

This proves the proposition.

Let k be a perfect field and let $R = k[x, y]/(x^2 + y^2-1)$ be the coordinate ring of the circle. If $i = \sqrt{-1} \in k$ then $R \cong k[t, t^{-1}]$ by $x + iy \to t$, $x - iy \to t^{-1}$. By the previous proposition $Br(R)=Br(k)\oplus\hat{G}$. In particular, if k is a finite field containing -1 then there are two 'points at infinity' on the circle over k by Theorem 2.2. If $i \notin k$ then any element in $Br(R)$ is split by $N[x, y]/(x^2 + y^2 - 1)$ where N is a normal extension of k containing $k[i]$. Since $i \in N$, we have $N[x, y]/(x^2 + y^2 - 1) \cong N[t, t^{-1}]$. Let $G(N/k)$ and $G(N/k[i])$ be the Galois groups of N over k and $k[i]$. Then

$$\begin{aligned}
Br(N[t, t^{-1}]/R) &= H^2(G(N/k), U(N[t, t^{-1}])) \\
&= H^2(G(N/k), U(N)) \times H^2(G(N/k), \langle t \rangle) \\
&= Br(N/k) \times H^2(G(N/k), Z)
\end{aligned}$$

where if $\sigma \in G(N/k[i])$ then σ acts trivially on Z and if $\sigma \notin G(N/k[i])$ then $\sigma \cdot n = -n$ for all $n \in Z$. In particular, if $G(N/k)$ is abelian one can check that with this action on Z one has $H^2(G(N/k), Z) = \{1\}$ so $Br(N[t, t^{-1}]/R) = Br(N/k)$. Thus if k is finite and $i \notin k$ then $Br(R) = (0)$ and there is one closed point at infinity on the circle. If we let k be the p-adic completion of Q at the prime 11 then $i \notin k$ and $x^2 - x + 1$ is irreducible in $k[x]$ so if N is the splitting field of $x^2 - x + 1$ then $i \in N$ and $G(N/k) = S_3$. In this case

$$Br(N[t, t^{-1}]/R) = Br(N/k) \times H^2(S_3, Z)$$

where S_3 is the full symmetric group and where the transpositions in

S_3 invert the elements in Z. In this case $H^2(S_3, Z) = Z/(3)$ so

$$Br(N[t, t^{-1}]/R) = Br(N/k) \oplus Z/(3).$$

On the other hand, if T is the abelian closure of k then

$$Br(T[t, t^{-1}]/R) = Br(T/k).$$

Even for p-adic fields one does not expect then a formulae for the Brauer group of a curve as given in Theorem 2.1 and Theorem 2.2 for finite fields or the field of real numbers.

Now we consider the Brauer group of a real surface. Let $R = \mathbb{R}[x, y, z]/(xyz - 1)$. Then one can check that $R = \mathbb{R}[t,t^{-1},u,u^{-1}]$ for indeterminates t, u by $x \to t$, $y \to u$ and $z \to t^{-1}u^{-1}$. Let $S = \mathbb{C} \otimes_{\mathbb{R}} R$, and C_2 the Galois group of $\mathbb{C}$ over $\mathbb{R}$ where $\mathbb{R}$ = real numbers and $\mathbb{C}$ = complex numbers. Then

$$\begin{aligned} Br(S/R) &= H^2(C_2, U(S)) \\ &= H^2(C_2, U(\mathbb{C})) \times H^2(C_2, \langle t\rangle) \times H^2(C_2, \langle u\rangle) \end{aligned}$$

where C_2 acts trivially on $\langle t\rangle$ and $\langle u\rangle$.

Thus $Br(S/R) = Z/(2) \oplus Z/(2) \oplus Z/(2)$; however, the real surface $xyz - 1$ over $\mathbb{R}$ has six connected components in the strong topology. It is unlikely that any analog of Theorem 2.1 is possible for real affine surfaces. The part of the Brauer group split by S is studied in [6]. If X is a real projective surface, is the subgroup of Br(X) split by $\mathbb{C}$ a two group whose rank is the number of connected components of X?

REFERENCES

[1] L. N. Childs,"Mayer-Vietoris sequences and Brauer groups of non-normal domains", Trans-Amer. Math. Soc. 196, 51-67, 1974.

[2] F. R. DeMeyer, "The Brauer group of a ring modulo an ideal", Rocky Mtn. J. of Math. (to appear).

[3] F. R. DeMeyer and M. A. Knus, "The Brauer group of a real curve", P.A.M.S. (to appear).

[4] A. Grothendieck, "Le groupe de Brauer I, II, III, in: Dix exposés sur la cohomologie des schèmas", Paris, Masson et Amsterdam, North Holland, 1968, 46-188.

[5] M. A. Knus, M. Ojanguren, "A Mayer-Vietoris sequence for the Brauer group", J. of Pure and Applied Algebra, 5(1974), 345-360.

[6] Yu, Manin, "Cubic forms", North Holland Math. Library Vol. 4, 1974.

[7] E. Witt, "Zerlegung reeler algebraischer Funktionen in Quadrate Shiefkörper über reelem Funktionenkorper", J. für Math. 171, 4-11, (1934).

On Brauer groups in characteristic p

M. A. Knus, M. Ojanguren and D. J. Saltman*

1. Introduction

This paper is a joint and improved version of the talks by the first and the third author. We show that some general results on Brauer groups of fields of prime characteristic are valid for rings. For example Br(R) is p-divisible for any ring of prime characteristic p and Br(R) ⟶ Br(K) is surjective for any extension K of R such that for each element $x \in K$ a q-th power x^q lies in R, q a power of p.

These results were proved independently by the first two (M.A.K., M.O.) and the third (D.S.) author. The proofs of the surjectivity were different but both were using in some way Berkson's theorem. After the conference the third author (D.S.) found a very direct proof. This proof is presented here. A useful tool is the isomorphism Br(R) ⟶ Br(R/I) for any nilpotent ideal I. This result has already been proved by different authors (Hoobler, Giraud, DeMeyer, see the references). In § 2 we give a short proof of a slightly more general result which essentially follows Grothendieck [Gr]. Then we study the behavior of the Brauer group under the Frobenius map for rings of characteristic p. In § 4 come the results mentioned at the beginning. Finally we give different applications. The first two authors consider polynomial rings in characteristic p. They show that such rings in more than one variable over finite fields are infinite countable direct sums of copies of $\mathbb{Z}(p^\infty)$ and that the kernel of Br(K[T]) ⟶ Br(K) is an infinite direct sum of copies of $\mathbb{Z}(p^\infty)$ if K is not perfect. The third author first generalizes known results about extensions of derivatives in separable algebras (for this R does not need characteristic p).

*Research (D.S.) partially supported by NSF grant MPS 72-04643.

Then he gives an explicit set of generators for the subgroup of $Br(R)$ of all elements of exponent p, R any ring of characteristic p. As a consequence he shows that for any ideal I of R the p-primary part of $Br(R) \longrightarrow Br(R/I)$ is a split epimorphism. Furthermore the kernel of $Br(R) \longrightarrow Br(R/I)$ is always p-divisible. The notes [DI], [KO] or [OS] are references for unexplained notions. The symbol R will always denote a commutative ring and tensor products without subscripts mean tensor products over R.

2. Lifting algebras

Let R be a commutative ring and I an ideal of R. For any R-module or R-homomorphism we denote by a bar its tensor product with $\bar{R} = R/I$.

We refer to Cartan-Eilenberg, [CE], Chap. IX, for the definition and properties of Hochschild cohomology.

Theorem 2.1: Let A be a finitely generated projective R-module. Suppose there is a multiplication $\bar{\lambda}: \bar{A} \otimes \bar{A} \longrightarrow \bar{A}$ and an element $\bar{1} \in \bar{A}$ making $\bar{A}$ an associative unitary $\bar{R}$ algebra. If R is I-adically complete and the Hochschild dimension of $\bar{A}$ is not greater than two, there is a multiplication $\lambda: A \otimes A \longrightarrow A$ and an element $1 \in A$, which lift $\bar{\lambda}$ and $\bar{1}$ correspondingly, and which make A an associative unitary R algebra. This multiplication and unit are unique up to isomorphism if the Hochschild dimension of $\bar{A}$ is zero or one.

Proof: We will deduce this theorem from the special case of a nilpotent ideal I. Clearly, by induction on the index of nilpotency of I, it is enough to consider the case $I^2 = 0$. Since $A \otimes A$ is R-projective the multiplication $\lambda: \bar{A} \otimes \bar{A} \longrightarrow \bar{A}$ can be lifted to an R-homomorphism $\nu: A \otimes A \longrightarrow A$. We define on IA a structure of $\bar{A}$-bimodule by setting $\bar{a}x = \nu(a \otimes x)$ and $x\bar{a} = \nu(x \otimes a)$ for any x in

IA. It follows from $I^2 = 0$ that, for a fixed $\bar{a}$, $\bar{a}x$ and $x\bar{a}$ are independent of a and that this structure of bimodule depends on λ but not on the choice of ν. The multiplication defined by ν is R-bilinear but may fail to be associative. Therefore we look at $F(a,b,c) = \nu(\nu(a\otimes b)\otimes c) - \nu(a\otimes\nu(b\otimes c))$. Since $\bar{\nu} = \lambda$ is associative, $F(a,b,c)$ is in IA and since $I^2 = 0$ it only depends on $\bar{a},\bar{b},\bar{c}$. A short computation shows that F is a Hochschild 3-cocycle of $\bar{A}$ with values in IA, hence a coboundary: there is an $\bar{R}$-linear map $\theta'\colon \bar{A}\otimes\bar{A} \longrightarrow IA$ such that $F(\bar{a},\bar{b},\bar{c}) = \bar{a}\theta'(\bar{b}\otimes\bar{c}) - \theta'(\overline{ab}\otimes\bar{c}) + \theta'(\bar{a}\otimes\overline{bc}) - \theta'(\bar{a}\otimes\bar{b})\bar{c}$. We define a new lift μ of λ by $\mu = \nu + \theta$ where θ is the composite of θ' with the canonical projection of $A\otimes A$ onto $\bar{A}\otimes\bar{A}$. A computation which we again omit shows that μ is associative. If 1 is the unit of $\bar{A}$ there is an idempotent e in A such that $\bar{e} = 1$. It is easy to see that $eA = Ae = A$, hence e is the unit of A.

We now prove the first part of the theorem. Suppose that R is I-adically complete, i.e. $R = \varprojlim R/I^n$. Then $A = \varprojlim A/I^nA$ as R-module. For each n we can define an R-algebra structure on A/I^nA such that, for $n \geq m$ the canonical maps $A/I^nA \longrightarrow A/I^mA$ are R-algebra homomorphisms. Then $\varprojlim A/I^nA$ is an R-algebra which lifts $\bar{A}$.

To show uniqueness we assume that $\bar{A}$ has Hochschild dimension at most one and start again with the case $I^2 = 0$. Let μ and ν be two associative multiplications with unit on A inducing the same multiplication λ on $\bar{A}$. Put $F(a,b) = \mu(a\otimes b) - \nu(a\otimes b)$ for any a,b in A. Since $\bar{\mu} = \bar{\nu}$, $F(a,b)$ is in IA and because of the assumption $I^2 = 0$ it only depends on $\bar{a},\bar{b}$. It is easily seen that F is a Hochschild 2-cocycle of $\bar{A}$ with values in IA. Hence it is a coboundary: $F(\bar{a},\bar{b}) = \bar{a}\theta'(\bar{b}) - \theta'(\overline{ab}) + \theta'(\bar{a})\bar{b}$ for some $\bar{R}$-linear map $\theta'\colon \bar{A} \longrightarrow IA$. Let $\theta\colon A \longrightarrow IA$ be the composite of θ' with the canonical projection of A onto $\bar{A}$ and define $f: A \longrightarrow A$ by $f(a) = a + \theta(a)$. An easy computation shows that $f(\mu(a \otimes b)) = \nu(f(a) \otimes f(b))$ for any a,b in A. Let e_μ be the unit of μ and e_ν that of ν. Then $f(\mu(e_\mu \otimes e_\mu)) = f(e_\mu) = \nu(f(e_\mu) \otimes f(e_\mu))$. This shows that $f(e_\mu)$ is an idempotent for ν, but since it lifts the unit of $\bar{A}$ it must be e_ν. We conclude that f is an isomorphism between the two lifts of λ, which induces the identity on $\bar{A}$. The general case follows by a limit argument as before. Q.E.D.

The following theorems are consequences of Theorem (2.1).

Theorem 2.2: Let R be I-adically complete. Then $\otimes R/I$ induces a bijection from the set of isomorphism classes of Azumaya algebras over R onto the set of isomorphism classes of Azumaya algebras over $\bar{R}$.

Proof: If $\bar{A}$ is Azumaya over $\bar{R}$ then by, for example, [B] p. 10, $\bar{A}$ lifts to a faithful projective R-module A. By (2.1), A can be taken to be an R-algebra. Since $\bar{A} \otimes \bar{A}^0 \cong \mathrm{End}_{\bar{R}}(\bar{A})$, by the uniqueness part of (2.1), $A \otimes A^0 \cong \mathrm{End}_R(A)$, and thus A is Azumaya over R. Again by the uniqueness assertion in (2.1), A is the only preimage, up to isomorphism, of $\bar{A}$. Q.E.D.

The following theorem is now immediate.

Theorem 2.3: Let R be I-adically complete. Then R and R/I have the same Brauer group.

Theorem 2.4: Let $\bar{R} = R/I$, where I is a nil ideal. Then the canonical projection induces an isomorphism $\mathrm{Br}(R) \xrightarrow{\sim} \mathrm{Br}(\bar{R})$.

Proof: Write $R = \varinjlim R_i$ where the limit is taken over all noetherian subrings of R. The induced map $\varinjlim \mathrm{Br}(R_i) \longrightarrow \mathrm{Br}(R)$ is surjective since any Azumaya over R is already defined over a noetherian subring of R (see for example [KO] Prop. III 5.7). It is also clearly injective since an algebra A_0 defined over some noetherian subring R_0 of R which is trivial over R is already trivial over some finitely generated extension R_1 of R_0. Therefore we have

$$(2.5) \qquad \varinjlim \mathrm{Br}(R_i) \cong \mathrm{Br}(R).$$

Call $\bar{R}_i = R_i/R_i \cap I$, then $\bar{R} = \varinjlim (\bar{R}_i)$ and by (2.3) $\mathrm{Br}(R_i) \cong \mathrm{Br}(\bar{R}_i)$ which proves the wanted isomorphism. Q.E.D.

Remark 2.6: M. Cipolla [C] has recently generalized Th. (2.1) to any Hensel pair (R,I).

3. The Brauer group and purely inseparable extensions

If R is a commutative ring of characteristic p and if q is a power of p then the map $\pi: R \longrightarrow R$ defined by $\pi(x) = x^q$ is a ring homomorphism. We will study the induced map $\pi^*: \mathrm{Br}(R) \longrightarrow \mathrm{Br}(R)$ and show that π^* is the multiplication by q map on the abelian group $\mathrm{Br}(R)$.

The major tool in this study will be the embedding of the Brauer group into étale cohomology. Let us recall the definitions. A commutative R algebra S is an étale covering of R if 1) S is faithfully flat, 2) S is finitely presented (as an R algebra), and 3) S is locally separable (that is, if Q is a prime ideal of S and $P = Q \cap R$ then S_Q/PS_Q is separable over R_P/PR_P. As in, for example [OS] p. 153, define

$$(3.1) \qquad H^n(R) = \varinjlim H^n(S/R,U).$$

Here $H^n(S/R,U)$ is the Amitsur cohomology group with respect to the units functor U and the direct limit is taken over all isomorphism classes of étale coverings of R. This direct limit is well defined because of the following result.

Theorem 3.2: Suppose S and S' are commutative R and R' algebras respectively and suppose that $f_1, f_2 \colon S \longrightarrow S'$ are two ring homomorphisms such that $f_i(R) \subseteq R'$ and $f_1 = f_2$ on R. Then the induced maps $f_1^*, f_2^* \colon H^n(S/R,F) \longrightarrow H^n(S'/R',F)$ are equal ($n = 0,1$ if F is a functor into non abelian groups and $n \geq 0$ if F is into abelian groups).

Proof: In, for example, [KO] p. 123 or [OS] p. 151 this result is proved in the case $R = R'$ and the f_i are R algebra homomorphisms. The proof of the above is, in fact, word for word the same. Q.E.D.

Now we recall how the embedding mentioned above, written $\lambda\colon \mathrm{Br}(R) \longrightarrow H^2(R)$, is constructed. Let A be an Azumaya R-algebra. Then there exists an étale covering S/R such that

$$(3.3) \qquad \varphi\colon\ S \otimes_R A \cong \mathrm{End}_S(P)$$

for some finitely generated projective and faithful S module P. Notice that if A has constant rank over R, one can always choose S in such a way that P becomes a free S module (e.g. [KO] p. 105).

For any homomorphism of modules $\beta\colon M_1 \otimes_R \cdots \otimes_R M_k \longrightarrow N_1 \otimes_R \cdots \otimes_R N_m$ denote by β_i the homomorphism obtained by tensoring β with the identity of S in the i-th position. Consider now the $S \otimes S$ isomorphism ψ defined by the commutativity of

$$(3.4) \qquad \begin{array}{ccc} S \otimes S \otimes A & \xrightarrow{\ \varphi_1\ } & \mathrm{End}_{S\otimes S}(S \otimes P) \\ \Big\downarrow{\scriptstyle \tau_1} & & \Big\downarrow{\scriptstyle \psi} \\ S \otimes A \otimes S & \xrightarrow{\ \varphi_2\ } & \mathrm{End}_{S\otimes S}(P \otimes S) \end{array}$$

where τ is the switch. Then $\psi_2 = \psi_3\psi_1$, i.e. ψ is a faithfully flat descent datum from S to R for $\mathrm{End}_S(P)$ (and the descended algebra is, of course, A!) (see, e.g., [KO] p. 38). Even if

Skolem-Noether is not true in general, one can modify S in such a way that ϕ is induced by an $S \otimes S$ isomorphism of modules (e.g. [OS] p. 157)

(3.5) $$\alpha: \; S \otimes P \cong P \otimes S.$$

Since $\phi_2 = \phi_3\phi_1$, there is a unit $u \in S \otimes S \otimes S$ such that

(3.6) $$u\alpha_2 = \alpha_3\alpha_1.$$

One easily verifies that u is a 2-cocycle in Amitsur cohomology, which has an image $[u] \in H^2(R)$. We define $\lambda(A) = [u]$.

Theorem 3.7: Suppose R and R' are commutative rings and $f: R \longrightarrow R'$ is a ring homomorphism such that for any étale covering S of R, f extends to a homomorphism (also called f) $f: S \longrightarrow S'$ where $S' = S \otimes_R R'$. Then f induces a homomorphism $f^*: H^2(R) \longrightarrow H^2(R')$ independent of the choices of extensions and the following diagram commutes

$$\begin{array}{ccc} Br(R) & \xrightarrow{\lambda} & H^2(R) \\ {\scriptstyle f^*}\downarrow & & \downarrow{\scriptstyle f^*} \\ Br(R') & \xrightarrow{\lambda} & H^2(R') \end{array} .$$

Proof: As usual, R' is an R-algebra through f. By [OS] p. 149, $S' = S \otimes_R R'$ is an etale covering of R'. An extension of f induces a homomorphism $f^*: H^2(S/R,U) \longrightarrow H^2(S'/R',U)$ which, by Theorem 3.2, is independent of the extension chosen. Also by 3.2, these maps f^* commute with the direct limit so that $f^*: H^2(R) \longrightarrow H^2(R')$ is induced. We need the following lemma

Lemma 3.8: Suppose A and A' are Azumaya algebras over R and R' respectively, and suppose A and A' are of constant and equal rank. If there is a f-semilinear ring homomorphism $f': A \longrightarrow A'$ then $A' \cong A \otimes_R R'$.

Proof: It is clear that $f'(A)$ is Azumaya over $f(R) = f'(R) \subseteq R'$. Since $f'(A) \otimes_R R'$ is Azumaya over R, and since any non-zero ideal of an Azumaya algebra has a non-zero intersection with the center ([KO] p. 95), it is easy to see that the multiplication map $f'(A) \otimes_R R' \longrightarrow f'(A)R'$ is an isomorphism. Thus $f'(A)R'$ is Azumaya over R'. Checking ranks and using the double centralizer theorem ([DI] p. 57), we get that $f'(A)R' = A'$. The map $A \otimes_R R' \longrightarrow A'$ given by $a \otimes r' \longmapsto f'(a)r'$ is also easily seen to be an isomorphism. Q.E.D.

Let us finish the main proof. Suppose A is Azumaya over R. Replacing A by some equivalent algebra, we can suppose that A is of constant rank n^2. Choose S an étale covering of R such that in the splitting (3.3) P is a free module, that is, $P = S^n$. Then we can consider the isomorphism α of (3.5) as the inner autormorphism of $M_n(S \otimes S)$ determined by α. By [KO] p. 38 A can be identified with the following subring of $M_n(S)$, $A = \{x \in M_n(S) \mid \psi(1 \otimes x) = x \otimes 1\}$. Choose an extension $f\colon S \longrightarrow S'$. f induces a map $f\colon M_n(S \otimes S) \longrightarrow M_n(S' \otimes S')$ and call $\alpha^f = f(\alpha)$. α^f induces, by conjugation, an automorphism ψ^f of $M_n(S' \otimes S')$. Since $\psi_2 = \psi_3\psi_1$, $\psi_2^f = \psi_3^f\psi_1^f$. Thus ψ^f determines, by faithfully flat descent, an algebra $A^f = \{x \in M_n(S') \mid \psi^f(1 \otimes x) = x \otimes 1\}$. A^f is Azumaya over R' of rank n^2. The induced map $f\colon M_n(S) \longrightarrow M_n(S')$ restricts to a map $f\colon A \longrightarrow A^f$ and so by Lemma 3.8, $A^f \cong A \otimes_R R'$. If ψ determines the 2-cocycle $u \in H^2(S/R,U)$ then clearly ψ^f will determine $u^f = f^*(u) \in H^2(S'/R',U)$. This completes the proof. Q.E.D.

Finally, the point of this section, and the relevance to the characteristic p Brauer group, is embodied in the following.

Theorem 3.9: Let R be a commutative ring of characteristic p, q a power of p, and $\pi\colon R \longrightarrow R$ the map $\pi(x) = x^q$. Then the induced map π^*: $Br(R) \longrightarrow Br(R)$ is the multiplication by q map.

Proof: π clearly extends to any étale extension S of R by setting $\pi(s) = s^q$. Thus, by Theorem 3.7, the following commutes

$$\begin{array}{ccc} Br(R) & \xrightarrow{\lambda} & H^2(R) \\ \pi^* \downarrow & & \downarrow \pi^* \\ Br(R) & \xrightarrow{\lambda} & H^2(R) . \end{array}$$

π^*: $H^2(R) \longrightarrow H^2(R)$ can trivially be seen to be the multiplication by q map and since λ is injective this does it. Q.E.D.

Corollary 3.10: Let K be any extension of R such that $K^q = R$, then $Br(K/R) = {}_qBr(R)$ = those elements of $Br(R)$ annihilated by q.

Proof: π: $R \longrightarrow R$ can be factored as $\pi' \circ i$, where i: $R \longrightarrow K$ is the inclusion and π': $K \longrightarrow R$ is the map $\pi'(k) = k^q \in R$. Thus the induced maps on the Brauer groups satisfy $\pi^* = (\pi')^* \circ i^*$. π' itself factors as $K \longrightarrow K/M \cong R$ where M is the kernel of π'. M is an nil ideal since $\pi'(k) = 0$ means $k^q = 0$. Thus by Theorem 2.4 and the functoriality of the Brauer group, $(\pi')^*$: $Br(K) \longrightarrow Br(R)$ is an isomorphism. Hence the kernel of π^* is the kernel of i^* and this is the result. Q.E.D.

4. On a generalization of a theorem of Albert-Hochschild and the p-divisibility of the Brauer group

The following is an easy extension of Theorem 5.5 of [DI] p. 64. We will omit the proof.

Theorem 4.1: Suppose A is an Azumaya algebra over a commutative ring R of arbitrary characteristic and that A contains a commutative subalgebra S such that A is left projective over S. Then $B = A^S$ (= the centralizer of S in A) is Azumaya over S. Furthermore B and $A \otimes S$ are similar (i.e. equal in the Brauer group).

Now we come to the main business of this section. The following result is a generalization of Theorem 6.1 of [KO] p. 146 and of the Corollary to Theorem 6 of $[Y]_3$. We prove this result using the idea of extending derivations. This is the original approach of Hochschild, who proved the corresponding result when R is a field. The technique we use is similar to that of Hoechsmann, who also worked in the case R is a field.

Theorem 4.2: Let R be a ring of characteristic p and K an extension of R such that for each $k \in K$ there is a power q of p such that $k^q \in R$. Then the canonical map $Br(R) \longrightarrow Br(K)$ is surjective.

Proof: (Saltman) As a first step, let us show that we can assume K is a very special kind of purely inseparable extension. If A is Azumaya over K, then by Proposition 5.7 on p. 97 of [KO], there is a finitely generated $\mathbb{Z}$-subalgebra K_0 of K and an Azumaya algebra A_0 over K_0 such that $A_0 \otimes_{K_0} K \cong A$. By considering K_0R instead of K_0 we can take K_0 finitely generated as an R-algebra. Under this assumption, there is a power q of p such that $K_0{}^q \subseteq R$. It is easy to see that K_0 is an image of an R-algebra K' of the form

$$K' = R[x_1,x_2,\ldots,x_n]/(x_1{}^q-a_1,\ldots,x_n{}^q-a_n) .$$

Call I the kernel of the map $K' \longrightarrow K_0$. $I^q \subseteq R$ and so $I^q = 0$. By Theorem (2.4), it suffices to show $Br(R) \longrightarrow Br(K')$ is surjective. Since we can decompose $K' \supseteq R$ into a finite sequence of extensions of the form

$$C = R[x]/(x^p-a)$$

it suffices to prove our result for any such C. This will take some work.

There is a derivation $d': C \longrightarrow C$, which we will call special, and which is defined by the relations $d'(R) = 0$ and $d'(v) = 1$, where v is the image of x in C. Suppose now that A is an Azumaya algebra over C. By [Kn], d' extends to a derivation $d: A \longrightarrow A$. The existence of d is also guaranteed by Theorem 6.1 of this paper, which is proved independently of anything which precedes it.

It is well known ([J] p. 186) that $d^p = d \circ d \circ \dots \circ d$ (p times) is also a derivation of A. Since $d^p(C) = 0$, d^p is an inner derivation of A determined by, say, $u \in A$. We claim that we can choose u such that $d(u) = 0$. Firstly, $d(u) \in C$ since if $a \in A$, $d(u)a - ad(u) = [d(u),a] = d([u,a]) - [u,d(a)] = d(d^p(a)) - d^p(d(a)) = 0$. Write $d(u) = r_0 + r_1v + r_2v^2 + \dots + r_{p-1}v^{p-1}$. Then $0 = [u,u] = d^p(u) = d^{p-1}(d(u)) = (p-1)!\, r_{p-1}$ so $r_{p-1} = 0$. Thus it is easy to see that there is a $c \in C$ such that $d(c) = d(u)$. Replacing u by $u-c$ will prove our claim. So we assume $d(u) = 0$.

Form the differential polynomial ring $A[z,d]$. As a left A module, $A[z,d] \cong A \oplus Az \oplus Az^2 \oplus \dots$ and multiplication is defined using the relation $[z,a] = d(a)$ for all $a \in A$. The element z^p-a is central in $A[z,d]$. Define the algebra (A,d,u) to be

$$A[z,d]/(z^p-a).$$

Since z^p-a is central, it is not hard to see that (A,d,u) is free as a left A module with basis $1,z,\dots,z^{p-1}$.

<u>Remarks</u>: a) This construction coincides with the "differential extensions" of Hoechsmann [Ho], the "differential crossed products" of Knus and Ojanguren ([KO] p. 54) and the "Regular Lie algebra extensions" of Yuan ($[Y]_3$ p. 431) whenever the domains of definition intersect. In the work of Hoechsmann, C is assumed to be a field and

in the work of the other authors $A = C$ but C is more general.

b) Whenever we write (A,d,u) we will be assuming d is a derivation of A which induces a special derivation on C the center of A, d^p is the inner derivation of A induced by u, and $d(u) = 0$.

Proposition 4.3: (A,d,u) is Azumaya over R and A is the centralizer of C in (A,d,u).

Proof: If $t \in (A,d,u)$ centralizes C then $[v,t] = 0$. Call d_v the inner derivation determined by v. Write $t =$ $a_0 + a_1 z + a_2 z^2 + \dots + a_{p-1} z^{p-1}$. $0 = d_v{}^{p-1}(t) = (p-1)!\, a_{p-1}$. Thus $a_{p-1} = 0$. $0 = d_v{}^{p-2}(t) = (p-2)!\, a_{p-2}$, so $a_{p-2} = 0$. Proceeding in this way we get $a_{p-1} = a_{p-2} = \dots = a_1 = 0$. Hence $t \in A$. Clearly all of A centralizes C so A is the centralizer of C.

If Z is the center of (A,d,u) then by above, $Z \subseteq A$. Since the center of A is C, $Z \subseteq C$. Since Z centralizes z, $d(Z) = 0$ and so $Z \subseteq R$. Since R clearly centralizes (A,d,u), R is the center.

Finally, by [DI] p. 72, it will suffice to show that $(A,d,u)/M(A,d,u)$ is central simple over R/M for any maximal ideal M of R. Hence we can assume R is a field and must show (A,d,u) is a simple ring. C, however, may not be a field. Suppose $U \neq 0$ is an ideal of (A,d,u). If $w = a_0 + a_1 z + \dots + a_{p-1} z^{p-1} \in U$, then $d_v{}^{p-1}(w) = (p-1)!\, a_{p-1} = -a_{p-1} \in U$. Using $d_v{}^{p-2}$ we get that $a_{p-2} \in U$ and proceeding in this way we can show that $U =$ $U_0 + U_0 z + \dots + U_0 z^{p-1}$ where U_0 is an ideal of A. Since A is Azumaya over C, $U_0 = IA$ for some ideal I of C ([KO] p. 95). If $0 \neq c = r_0 + r_1 v + \dots + r_j v^j$ is in I and $r_j \neq 0$ then $d^j(c) = j!\, r_j \in I$ and $j!\, r_j$ is a unit. Thus $U = (A,d,u)$. Q.E.D.

Since A is Azumaya over C, A is projective as a C module. Since (A,d,u) is free as a left A module, (A,d,u) is projective as a left C module. Thus by Theorem 4.1, $(A,d,u) \otimes_R C$ is similar to $(A,d,u)^C = A$. That is, (A,d,u) represents a preimage of A in Br(R). This completes the proof of Theorem 4.2.

Corollary 4.4: For any ring of characteristic p, Br(R) is p-divisible.

Proof: Take any extension of R such that $K^p = R$ and note that the composition $Br(R) \longrightarrow Br(K) \longrightarrow Br(R)$ is the p-power map by 3.9 and is surjective by 4.2, since $Br(K) \longrightarrow Br(R)$ is an isomorphism by 2.4. Q.E.D.

5. Brauer groups of polynomial rings (M.A.K. and M.O.)

We first recall a formula due to Zelinsky for fields and Yuan $[Y]_1$ in general (the proof can also be found in [KO]). Let R be a ring of prime characteristic p and let S be an extension of R of the form $S = R[X]/(X^p - a)$. If d is the R-derivation of S such that $d(x) = 1$, where x is the class of X, then

$$(5.1) \qquad H^2(S/R) \cong R^+/\{d^{p-1}(c) + c^p, \quad c \in S\}$$

where R^+ is the additive group of R. For any extension S/R finitely generated projective as R-module there is a map

$$(5.2) \qquad \beta\colon \; H^2(S/R) \longrightarrow Br(S/R)$$

due to Rosenberg and Zelinsky (corresponding exactly to the cross-product construction in Galois cohomology). In our case, using that the kernel of the multiplication map $S \otimes_R S \longrightarrow S$ is nilpotent one can show that β is surjective. On the other side it follows from the Chase-Rosenberg sequence that β is injective if for example Pic $S = 0$. We now use Zelinsky's formula to compute a subgroup of $Br(R[T^{1/p}]/R[T])$ where $R[T]$ is the polynomial ring in one variable over R.

<u>Proposition 5.3</u>: Let R be a reduced (i.e. having no nilpotent elements) ring of prime characteristic p such that $Pic(R[T]) = 0$. Then $Br(R[T^{1/p}]/R[T])$ contains for each power $q = p^m$, $m = 1,2,\ldots,$ a subgroup isomorphic to R/R^q.

<u>Proof</u>: Let d be the $R[T]$-derivation of $R[T^{1/p}]$ such that $d(T^{1/p}) = 1$ and let $E = \{d^{p-1}(c) + c^p,\ c \in R[T^{1/p}]\}$. One verifies immediately that

$$E = \{x^pT^{\ell} \text{ for } \ell \neq np-1,\ x \in R;\ -yT^{n-1} + y^pT^{np-1},\ n = 1,2,\ldots,\ y \in R\}$$
$$= \{x^pT^{\ell} \text{ for } \ell \neq np-1,\ x \in R;\ -yT^{n-1} + y^qT^{nq-1},\ n = 1,2,\ldots,\ q = p^m,\ m = 1,2,\ldots,\ y \in R\}.$$

Let now $E_q = E \cap (R+RT^{q-1})$. We prove that

$$E_q = \{x^p,\ x \in R;\ -y + y^qT^{q-1},\ y \in R\}.$$

The inclusion $E_q \supset \{\ldots\}$ is clear. Let $a + bT^{q-1} \in E_q$. Since $-a + a^qT^{q-1} \in E_q$, $(b+a^q)T^{q-1} \in E_q$. But using that R is reduced it is easy to show that zT^{q-1} can be in E only if $z \in R^{pq}$. As we argue below, this gives $a+bT^{q-1} \in \{\ldots\}$. We now verify that the subgroup $R + RT^{q-1}/E_q$ of $R[T]/E$ is isomorphic to R/R^{pq}. It is clear for $q = 1$. Let now $q = p^m$, $m \geq 1$ and let $\rho\colon R \longrightarrow R/R^{pq}$ be the canonical projection. Define

$$\alpha\colon\ R + RT^{q-1} \longrightarrow R/R^{pq}$$

by $\alpha(a + bT^{q-1}) = \rho(a^q + b)$. Then clearly $\alpha(E_q) = 0$. If $\alpha(a + bT^{q-1}) = 0$, then $a^q + b = z^{pq}$ for some $z \in R$. Hence

$a + bT^{q-1} = a - (a^q - z^{pq})T^{q-1} = (a - z^p) - (a - z^p)^q T^{q-1} + z^p$ belongs to E_q. Q.E.D.

Let K be a field of prime characteristic p. It is well known that $Br(K) \cong Br(K[T])$ if and only if K is perfect, i.e. $K = K^p$. In general $Br(K[T]) = Br(K) \oplus Ker(Br(K[T]) \longrightarrow Br(K))$. Denote as usual

$$(5.4) \qquad Br'(K) = Ker(Br(K[T]) \longrightarrow Br(K)).$$

By (7.6) of [AG], $Br'(K)$ is a p-torsion group.

<u>Theorem 5.5</u>: The group $Br'(K)$ is an infinite direct sum of copies of $\mathbb{Z}(p^\infty)$ if K is non-perfect of characteristic p.

<u>Proof</u>: By (4.4) $Br(K)$ is p-divisible, hence $Br'(K)$, as a direct summand, is also p-divisible. Since we know that $Br'(K)$ is p-torsion, we obtain that $Br'(K)$ is divisible. Hence by the structure theorem for divisible groups ([K] p. 10), $Br'(K)$ is a direct sum of copies of $\mathbb{Z}(p^\infty)$. Clearly $Br(K[T^{1/p}]/K[T])$ is a subgroup of $Br'(K)$. By proposition 5.3 $Br(K[T^{1/p}]/K[T])$ is infinite over $\mathbb{Z}/p\mathbb{Z}$ (K/K^p is infinite if K is not perfect!). Hence the "rank" of $Br'(K)$ is infinite. Q.E.D.

We now compute the Brauer groups of polynomial rings over finite fields.

<u>Proposition 5.6</u>: Let K be a finite field of characteristic p. Then $Br(K[T_1,\dots,T_n])$ is a p-torsion group.

<u>Proof</u>: The proof goes by induction on the number n of variables. For $n = 0$, one has $Br(K) = 0$. Write $Br(K[T_1,\dots,T_n])$ as the direct sum of $Br(K[T_1,\dots,T_{n-1}])$ and $Ker(Br(K[T_1,\dots,T_{n-1}][T_n]) \longrightarrow Br(K[T_1,\dots,T_{n-1}]))$. This kernel is contained in the kernel of

$$Br(K(T_1,\ldots,T_{n-1})[T_n]) \longrightarrow Br(K(T_1,\ldots,T_{n-1})$$

which is p-torsion by (7.6) of [AG]. Q.E.D.

Theorem 5.7: Let K be a finite field of characteristic p. Then for $n > 1$, $Br(K[T_1,\ldots,T_n])$ is an infinite countable direct sum of copies of $Z(p^\infty)$.

Proof: The group $Br(K[T_1,\ldots,T_n])$ is p-torsion by (5.7) and p-divisible by (4.4). Hence using again the structure theorem for divisible groups it is a direct sum of copies of $\mathbb{Z}(p^\infty)$. We show first that the direct sum is not finite. By (3.10) we know that

$$(5.8) \qquad {}_qBr(K[T_1,\ldots,T_n]) = Br(K[T_1^{1/q},\ldots,T_n^{1/q}]/K[T_1,\ldots,T_n]).$$

Hence $Br(K[T_1,\ldots,T_{n-1},T_n^{1/p}]/K[T_1,\ldots,T_n])$ is a subgroup of ${}_pBr(K[T_1,\ldots,T_n])$ and by (5.4) is not finite. It remains to show that the direct sum is countable. By (6.6) and (5.8), there is a surjection

$$\bigoplus_{i=1}^{n} Br(K[T_1,\ldots,T_i^{1/p},\ldots,T_n]/K[T_1,\ldots,T_n]) \longrightarrow {}_pBr(K[T_1,\ldots,T_n]).$$

Therefore we see that ${}_pBr(K[T_1,\ldots,T_n])$ is countable by applying 5.1. Thus $Br(K[T_1,\ldots,T_n])$ must be a countable direct sum of copies of $\mathbb{Z}(p^\infty)$. Q.E.D.

6. Brauer groups of rings modulo an ideal (Saltman)

In this section we will examine a situation with no analogue in the case R is a field. We will derive some interesting properties of the map $Br(R) \longrightarrow Br(R/I)$ when I is any ideal of R. More specifically, we will show that the above map is surjective on elements of p-power exponent and has p-divisible kernel. The crucial result for our purpose is the description of the subgroup of $Br(R)$ of elements of exponent p (Theorem 6.7).

As a first step, we will prove a result about extending derivations in separable algebras. We do not need the full strength of this result, but we believe it is of independent interest. It's full generality may be new. In this result only, R need not have characteristic p.

<u>Theorem 6.1</u>: Let A be a separable algebra over a commutative ring R and B a subalgebra such that A is projective as a left B-module. Let M be an A/R bimodule and $d': B \longrightarrow M$ a derivation such that $d'(R) \subseteq M^A = \{m \in M \mid am = ma \text{ for all } a \in A\}$. Then d' extends to a derivation $d: A \longrightarrow M$.

<u>Remark</u>: If B is a maximal commutative subalgebra of A, this result intersects one proved by Yuan in ($[Y]_3$ p. 430). In [Kn] and [BK] this result is proved in the case $B = R$. Our proof here is a refinement of the one from [BK].

<u>Proof</u>: Form the split ring extension X, of A by M. That is, the underlying additive group of X is $A \oplus M$ and multiplication is defined by $(a,m)(a',m') = (aa',am'+ma')$. X is an associative ring with unit (1,0). Identify A and M with $A \oplus 0$ and $0 \oplus M$ respectively and note that M is an ideal of X with $M^2 = 0$. There is a ring homomorphism $\pi: X \longrightarrow A$ defined by $\pi(a,m) = a$ and this homomorphism has kernel M. Use the derivation d' to give X a left B-module structure as follows: $b\cdot(a,m) = (ba,d'(b)a + bm)$. It is obvious that π is also a B-module morphism.

One can easily see that there is a one-to-one correspondence between derivations $d: A \longrightarrow M$ and ring homomorphisms $u: A \longrightarrow X$ such that $\pi \circ u = id_A$ = the identity map on A. This correspondence is given by $d \longmapsto u$ where $u(a) = (a,d(a))$. I claim that a derivation d extends d' if and only if the corresponding u is a B-module morphism. This is because $u(ba) = (ba,d(ba)) = (ba,d(b)a + bd(a))$ and $b\cdot u(a) = b\cdot(a,d(a)) =$

$(ba,d'(b)a + bd(a))$. Thus to finish this proof it suffices to find a B-linear ring homomorphism $u: A \longrightarrow X$ such that $\pi \circ u = id_A$.

Since A is B-projective, there is a B-linear $u: A \longrightarrow X$ such that $\pi \circ u = id_A$. We can assume $u(1) = (1,0)$ since we can replace u by $u'(a) = u(a) + a((1,0) - u(1))$. It remains to show that we can choose u to be a ring homomorphism. For any B-linear $u: A \longrightarrow X$ such that $u(1) = (1,0)$ and $\pi \quad u = id_A$, define $f_u: A \otimes_R A \longrightarrow M$ by $f_u(a \otimes a') = u(a)u(a') - u(aa')$. By, for example, [M] p. 285, f_u is a Hochschild 2-cocycle. If $h: A \longrightarrow M$ is B-linear and $h(1) = 0$ then $u' = u + h$ satisfies the same three conditions as u. We leave it to the reader to check that if $u' = u + h$ then $f_{u'} = f_u + \delta h$ where δ is the Hochschild coboundary map, that is, $\delta h(a,a') = ah(a') + h(a)a' - h(aa')$. Thus it suffices to construct, for any f_u, an h such that $f_u + \delta h = 0$.

If B were simply R, the result would be easy now since it is well known that the Hochschild cohomology group $H^2(A,M) = 0$. For our purposes, it is useful to explicitly construct the map h. A is separable over R, so there is an idempotent $e = \Sigma\, x_i \otimes y_i$ in the enveloping algebra $A \otimes_R A^o$ such that $\Sigma\, x_i y_i = 1$ and $\Sigma\, ax_i \otimes y_i = (a \otimes 1)e = (1 \otimes a)e = \Sigma\, x_i \otimes y_i a$ for all $a \in A$. Define $h: A \longrightarrow M$ by $h(a) = -\Sigma\, f_u(a,x_i)y_i$. We again leave it to the reader to check that h is B-linear, $h(1) = 0$, and $f_u + \delta h = 0$.

Q.E.D.

From now on, R will always be a commutative ring of characteristic p. Our goal at this point is to explicitly describe a set of Azumaya algebras which generate the subgroup of $Br(R)$ of all elements of exponent p. To this end it is necessary to show that every element of $Br(R)$ of p-power exponent is split by a "nice" purely inseparable extension. Let us introduce some notation and definitions. Denote by $R[a_1^{1/q},\dots,a_n^{1/q}]$ the R-algebra

$$R[x_1,\ldots,x_n]/(x_i{}^q - a_i)_{i=1,\ldots,n}$$

where q is a power of p. If $C = R[a^{1/p}]$ then we will, without further comment, refer to the image of x in C as v. C has a derivation, we call special, defined by the conditions $d(R) = 0$ and $d(v) = 1$.

Theorem 6.2: If A is Azumaya over R of exponent $q = p^e$ then there are $a_1,a_2,\ldots,a_m \in R$ such that A is split by $R[a_1^{1/q},\ldots,a_n^{1/q}]$.

Proof: Call $R^p = \{r^p \mid r \in R\}$, a subring of R, and suppose $\{a_i \mid i \in I\}$ generate R over R^p as an algebra. Define

$$C = R[\{x_i \mid i \in I\}]/(x_i^q - a_i)_{i\in I} \quad .$$

Then $C^q = R$. By Corollary 3.9, C splits A. Thus, as in Theorem 2.4, A is split by a finitely generated R subalgebra of C, C'. Adjoin to C' all the x_i appearing in generators of C' and call this subalgebra C''. C'' splits A and is of the required form.

Q.E.D.

If A is Azumaya over R and split by $C = R[a_1^{1/q},\ldots,a_n^{1/q}]$ then by [CR] p. 29, A is similar to an A' with A' containing C as a maximal commutative subalgebra of A' and A' left projective over C. When $C = R[a^{1/p}]$, we will show that we can describe A' completely. This is a corollary of the following theorem.

Theorem 6.3: Suppose A is an Azumaya algebra over R and suppose A contains a subalgebra $C = R[a^{1/p}]$ such that A is left projective over C. Then if $B = A^C$, the centralizer of C in A, there is a derivation d and an element u of B such that $A \cong (B,d,u)$.

Proof: By Theorem 4.1, B is Azumaya over C. Let d' be the special derivation of C with respect to v. By Theorem 6.1 above, d' extends to a derivation, d, of A. Since $d(C) \subseteq C$, $d(A^C) \subseteq A^C$. Since d is an R-derivation, there is a $z \in A$ such that d is the inner derivation determined by z, i.e., $d(a) = [z,a] = za - az$. By [J] p. 186, $d^p = d \circ d \circ d \circ \ldots \circ d$ (p times) is the inner derivation determined by $u = z^p$. Since $d^p(C) = 0$, $u \in A^C = B$. Call A' the subalgebra of A generated by B and z. Since $d(u) = [z,u] = [z,z^p] = 0$, there is a surjective R-algebra homomorphism

$$\varphi: \ (B,d,u) \longrightarrow A'.$$

(B,d,u) is Azumaya over R so kernel $\varphi = I \cdot (B,d,u)$ where $I \lhd R$ is an ideal of R ([KO], p. 95). But $r \longrightarrow r \cdot 1$ is an injection into A' so $I = 0$ and thus φ is an isomorphism. Hence A' is Azumaya over R. $(A)^{A'} \subseteq A^B = C$ and so $A^{A'}$ is commutative. By the double centralizer theorem ([DI] p. 57) $A^{A'} = R$ and $A' = A$.

Q.E.D.

The particular case $B = C$ is of interest to us. If $C = R[a^{1/p}]$, $u = b \in R$, and $d': C \longrightarrow C$ is the special derivation mentioned above, then (C,d',b) is well defined and isomorphic to

$$(1) \qquad R\langle x,y\rangle / (x^p - a, y^p - b, yx - xy - 1)$$

where $R\langle x,y\rangle$ is the free noncommutative algebra over R generated by x and y. Call $(a,b)_R$ the algebra (1) defined above. We will uniformly use the letters v and w for the images of x and y in $(a,b)_R$. The following is a special case of Theorem 6.3.

Corollary 6.4: If A is Azumaya over R containing $C = R[a^{1/p}]$ as a maximal commutative subalgebra and A is left projective over C then there is a $b \in R$ such that $A \cong (a,b)_R$.

These algebras $(a,b)_R$ satisfy the following elementary properties. Since the proofs are relatively easy, we will omit them.

Proposition 6.5:

a) $(a,b)_R$ is free as an R-module with basis $\{v^i w^j \mid 0 \leq i,j < p\}$ and if S is a commutative R-algebra, $(a,b)_R \otimes S \cong (a,b)_S$.

b) $(a,b)_R \cong M_p(R)$ if a is a p-th power in R.

c) $(a,b) \cong (-b,a)$; $(a,b)^o \cong (a,-b) \cong (-a,b)$.

d) Let v and w be the elements of $(a,b)_R$ mentioned above. If $c \in R[v]$ (the ring generated by R and v), then

$$(a,b)_R \cong (a, b+c^p+d_w{}^{p-1}(c))$$

where d_w is the inner derivation determined by w. Similarly, if $c \in R[w]$,

$$(a,b)_R \cong (a+c^p+d_v{}^{p-1}(c), b)_R.$$

e) $(a,b) \otimes (a,b') \cong (a,b+b') \otimes (0,b)$ so $(a,b) \otimes (a,b')$ is similar to $(a,b+b')$.

Call $Br(R)_p$ the subgroup of $Br(R)$ of all elements of p-power exponent. We study the subgroup $Br(R)_p$ by first studying its subgroup consisting of all elements of exponent p. We claim that this last subgroup is generated by the $(a,b)_R$'s. This is a consequence of the following.

Lemma 6.6: If A is Azumaya over R and split by $C = R[a_1^{1/p},\ldots,a_n^{1/p}]$ then A is similar to $A_1 \otimes A_2 \otimes \ldots \otimes A_n$ where each A_i is split by $R[a_i{}^{1/p}]$.

Remark: Something stronger is, in fact, true. Let A be Azumaya over R and let it contain a subalgebra $C = R[a_1^{1/q_1},\ldots,a_n^{1/q_n}]$, where q_i is a power of p, such that C is maximal commutative in A and A is left projective over C. Then $A \cong A_1 \otimes \dots \otimes A_n$ where A_i contains $C_i = R[a_i^{1/q_i}]$ as a maximal commutative subalgebra and A_i is left C_i projective. This result is proved in $[Y]_2$ using cohomological methods. It may also be done more directly, as in [S]. For the sake of completeness, we will give an easy proof of the simpler fact stated above.

<u>Proof</u>: We perform the proof of this lemma using induction on n. If $n = 1$, $C = R[a_1^{1/p}]$ and the result is trivial. Assume the result for $n-1$. Call $C' = R[a_1^{1/p},\dots,a_{n-1}^{1/p}]$. Then $A \otimes C'$ is split by $C'[a_n^{1/p}]$ and so $A \otimes C' \cong (a_n,b)_{C'}$, for some $b \in C'$. Consider $f = bv^{p-1}$, where $v,w \in (a_n,b)$ have the obvious interpretation. Note that $f^p + d_w^{p-1}(f) = a_n^{p-1}b^p + (-b)$ so by Proposition 6.5 d), $(a_n,b)_{C'} \cong (a_n, b + a_n^{p-1}b^p - b)_{C'} \cong (a_n, a_n^{p-1}b^p)$. Now $a_n^{p-1}b^p \in R$ so without loss of generality, $A \otimes C'$ is similar to $(a_n,b)_{C'}$ with $b \in R$. But then $A \otimes (a_n,b)_R^{\circ}$ is split by C' so we are done using the induction assumption. Q.E.D.

By Theorem 6.2 and the above lemma, any Azumaya algebra over R of exponent p is similar to a tensor product of algebras, A_i, each split by an extension $R[a_i^{1/p}]$. By Corollary 6.4 and the remark after 6.2, each A_i is similar to an algebra $(a_i,b_i)_R$. We have proved the following theorem.

<u>Theorem 6.7</u>: If R is a commutative ring of characteristic p then the subgroup of $Br(R)$ of all elements of exponent p is generated by the algebras $(a,b)_R$.

We can now state the main theorem of this section. Define $Br(R)_p$ to be those elements of $Br(R)$ of p-power exponent.

<u>Theorem 6.8</u>: Let R be a commutative ring of characteristic p and I an ideal of R. Then $Br(R)_p \longrightarrow Br(R/I)_p$ is a split epimor-

phism. Furthermore, the whole kernel of $Br(R) \longrightarrow Br(R/I)$ is p-divisible.

Proof: As promised, the first step of this proof is to examine the map $Br(R)_p \longrightarrow Br(R/I)_p$ restricted to elements of exponent p.

Lemma 6.9: If $I \triangleleft R$ is an ideal then $Br(R) \longrightarrow Br(R/I)$ is surjective when restricted to elements of exponent p.

Proof: By Theorem 6.7 it suffices to find a preimage in $Br(R)$ of any algebra $(\bar{a},\bar{b})_{R/I}$, $\bar{a},\bar{b} \in R/I$. But this simply entails finding preimages of $\bar{a}$ and $\bar{b}$ in R. Q.E.D.

To finish the proof of Theorem 6.8 it will suffice for us to observe two facts about abelian p-groups.

Lemma 6.10: If $f\colon M \longrightarrow N$ is a homomorphism of abelian p-groups, M is p-divisible and the image of f contains all $n \in N$ of exponent p then f is surjective. Furthermore, if every element $n \in N$ of exponent p has a preimage in M of the same exponent then the kernel of f is also p-divisible.

Proof: For the first part it suffices to note that the image of f is a divisible group and thus a direct summand of N. As for the second part, call K the kernel of f. If $m \in K$, there is an $m' \in M$ such that $pm' = m$. Thus $f(m')$ has exponent p and so there is a $k \in K$ such that $k+m'$ has exponent p. Thus $p(-k) = pm' = m$ and we are done. Q.E.D.

It is now easy to see that $Br(R)_p \longrightarrow Br(R/I)_p$ is an epimorphism with divisible kernel and thus splits. Finally, the kernel of $Br(R) \longrightarrow Br(R/I)$ is p-divisible since its p-primary part is p-divisible. This finishes the theorem and the section.

References

[AG] M. Auslander and O. Goldman, "The Brauer group of a commutative ring," Trans. Amer. Math. Soc. 97 (1960), 367-409.

[B] Hyman Bass, Algebraic K-Theory, Benjamin 1968.

[BK] Barr and Knus, "Extensions of derivations," Proc. Amer. Math. Soc. 28 (1971), 313-14.

[C] M. Cipolla, "Remarks on the lifting of algebras over Henselian pairs," (to appear, Math. Z.).

[CE] H. Cartan and S. Eilenberg, Homological Algebra, Princeton Math. Series 19, 1956.

[CR] Chase and Rosenberg, "Amitsur cohomology and the Brauer group," Mem. of Math. Soc. 52 (1963), 20-65.

[DeM] F. R. DeMeyer, "The Brauer group of a ring modulo an ideal," (to appear).

[DI] DeMeyer and Ingraham, Separable algebras over commutative rings, Springer L. N. 181, 1971.

[Gi] J. Giraud, Cohomologie non-abélienne, Springer Grundlehren 179, 1971.

[Gr] A. Grothendieck, "Le groupe de Brauer I," in Dix exposés sur la cohomologie des schémas, Paris: Masson, Amsterdam: North-Holland, 1968.

[H] R. Hoobler, A generalization of the Brauer group and Amitsur cohomology, Thesis, Berkeley, 1966.

[Ho] Hoechsmann, "Simple algebras and derivations," Trans. Amer. Math. Soc. 108 (1963), 1-12.

[J] Jacobson, Lie Algebras, Interscience Tracts in Pure and Appl. Math., No. 10, New York, 1962.

[K] I. Kaplansky, Infinite abelian groups, Univ. of Michigan Press, Ann Arbor, 1954.

[Kn] Knus, "Sur le theoreme de Skolem-Noether et sur les derivations de algèbras d'Azumaya," C. R. Acad. Sci. Paris Ser. A 270 (1970), 637-9.

[KO] M. A. Knus and M. Ojanguren, Theorie de la descente et algèbres d'Azumaya, Springer L. N. 389, 1974.

[M] Maclane, Homology, Springer, Berlin, 1971.

[OS] M. Orzech and Ch. Small, The Brauer group of commutative rings, Dekker L. N. 11, 1975.

[R] M. Raynaud, Anneaux locaux henséliens, Springer L. N. 169, 1970.

[S] Saltman, Azumaya algebras over rings of characteristic p, Thesis, Yale University, 1976.

$[Y]_1$ S. Yuan, "Brauer groups of local fields," Ann. of Math. (2) 82, (1965), 434-444.

$[Y]_2$ S. Yuan, "Brauer groups for inseparable fields," Ann. of Math., 96, (1974), 430-447.

$[Y]_3$ S. Yuan, "Central separable algebras with purely inseparable splitting rings of exponent one," Trans. Amer. Math. Soc. 153, Jan. 1971, p. 427.

A module approach to the Chase-Rosenberg-Zelinsky sequences

by

Gerald S. Garfinkel

Chase and Rosenberg [CR] have constructed a seven term exact sequence generalizing Hilbert's Theorem 90 and the classical cohomological description of the relative Brauer group. Their techniques involved using spectral sequences and Zarisky covers. In [RZ] Rosenberg and Zelinsky constructed an exact sequence which generalized part of the Skolem-Noether Theorem. In [G] by using non-abelian Amitsur cohomology sets and limits over a category $\underline{D}_0$ ($\simeq$ to the split Azumaya algebras) I constructed a sequence similar to that of Chase and Rosenberg. In the course of doing so I also gave a new proof of part of the Rosenberg-Zelinsky results. In my Conference talk I explained why I thought my techniques could be generalized to obtain the Chase-Rosenberg sequence itself. The promised generalizations are contained in this paper.

I found the Conference very stimulating and I am particularly grateful to four of the participants whose influences helped me produce this paper. To Bodo Pareigis for very helpful and encouraging discussions on this material; to Dan Zelinsky for his tireless efforts in organizing this Conference and these Proceedings; to S. A. Amitsur for inventing his cohomology theory [A] and for his work on Brauer splitting fields which indirectly first led me to consider sequences of non-abelian cohomology sets. Most of all this work was inspired by Ray Hoobler's thesis [H_1] which convinced me the Chase-Rosenberg sequence should be obtainable from non-abelian cohomology sets and that a method was needed to "patch together" cohomology sets into groups. I did this in my thesis via the category $\underline{D}_0$. (Hoobler later in [H_2] patched by means of sheaf cohomology. Incidently, three theses were somehow related to this work: Professor Amitsur informed me his Brauer field work came out of his thesis.)

§1. D^* and related categories.

R is always a commutative ring with 1 and S is a commutative R-algebra. $\underline{M}(S)$ is the full category of faithfully projective (i.e. finitely generated, faithful and projective) S-modules. $\underline{PIC}(S)$ the subcategory of rank one projectives and $\mathrm{Pic}(S)$ the group of isomorphism classes $\langle J \rangle$ of J in $\underline{PIC}(S)$. $\underline{ALG}_S$ is the category of S-algebras (with 1). We define a new category $\underline{D}^* = \underline{D}(S/R)$ with the same objects as $\underline{M}(S)$ as follows.

For $M, N \in \underline{M}(S)$ let Premaps (M, N) be all pairs (α, A) with $A \in \underline{M}(R)$ and $\alpha : A \otimes M \to N$ an S-isomorphism. We define an equivalence relation by $(\alpha, A) \sim (\beta, B)$ if there is $\tau : A \to B$ an R-isomorphism with

$$\begin{array}{ccc} A \otimes M & \xrightarrow{\quad\alpha\quad} & N \\ {\scriptstyle \tau \otimes M}\downarrow & & \| \\ B \otimes M & \xrightarrow{\quad\beta\quad} & N \end{array}$$

commutative and denote the equivalence class of (α, A) by $\langle \alpha, A \rangle$. We then define $\underline{D}^*(M, N) = \text{Premaps}\ (M, N)/\sim$. Since $\underline{M}(R)$ has a small skeleton $\underline{D}^*(M, N)$ is a set. We define composition on the premap level by $(\beta, B) \cdot (\alpha, A) = (\beta \cdot B \otimes \alpha, B \otimes A)$ for $(\alpha, A) : M \to N$ and $(\beta, B) : N \to P$. Composition is associative on the premap level and clearly induces a composition in $\underline{D}^*$. The identity map of M is $\langle \mu_M, R \rangle$ where $\mu_M : R \otimes M \to M$ is the canonical map.

We define other categories whose objects are those of $\underline{M}(S)$ as follows. $\underline{D}_S = \underline{D}(S/S)$. $\underline{PGL}_S(M, N) = \{\langle \alpha, S \rangle \in \underline{D}_S(M, N)\}$. $\underline{T}_S(M, N) = \{$iso classes $\langle J \rangle : J \otimes_S M \simeq N\}$. $\underline{GL}_S$ and $\underline{PIC}_S$ are the isomorphism categories of $\underline{M}(S)$ and $\underline{D}_S$ respectively - that is $\underline{GL}_S(M, N) = \mathrm{Iso}_S(M, N)$ and $\underline{PIC}_S(M, N) = \{\langle J \rangle : J \otimes_S M \simeq N$ and $J \in \underline{PIC}(S)\}$. Finally $\underline{U}_S(M, N) = U(S) =$ the units group of S.

Note $\underline{PGL}_S$ is a subcategory of $\underline{D}_S$ and $\underline{T}_S$ is a factor category of $\underline{D}_S$. In fact $\underline{T}_S$ is Bass' translation category [B] of $\underline{M}(S)$. $\underline{PIC}_S(M, N) = \underline{T}_S(M, N)$ if $\mathrm{rank}\ M = \mathrm{rank}\ N$. In a sense $\underline{U}_S$ operates on $\underline{GL}_S$ with orbits $\underline{PGL}_S$. In fact if we let $E_S(M) = \mathrm{Hom}_S(M, M)$, then

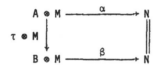

Theorem 1.1 *(Rosenberg and Zelinsky) There is a commutative diagram*

$$\begin{array}{ccccc}
\underline{PGL}_S(M, N) & \longrightarrow & \underline{D}_S(M, N) & \longrightarrow & \underline{T}_S(M, N) \\
\uparrow \theta' & & \uparrow \theta & & \| \\
\underline{GL}_S(M, N)/U(S) & \xrightarrow{\text{conj}} & \underline{ALG}_S(E_SM, E_SN) & \xrightarrow{RZ} & \underline{T}_S(M, N)
\end{array}$$

with $\theta(\rho) = \langle e, H\rangle$ *where* $H = \mathrm{Hom}_{E_SM}(M, N)$

and $e : H \otimes_S M \to N$ *is evaluation, conj is induced by conjugation and* RZ *is the Rosenberg-Zelinsky map in [RZ],* θ *and* θ' *are equivalences.*

Proof. The proof is essentially an application of Morita Theory. See [G] for details.

For $\underline{A} = \underline{D}, \underline{PGL}, \underline{GL}, \underline{U}, \underline{PIC}, \underline{ALG}_S E_S$ or $\underline{T}_S$, let $\underline{A}_S(M) = \underline{A}_S(M, M)$. Each $\underline{A}_S(M)$ is a group which is functorial in $\underline{D}^*$ as follows: for $\langle\alpha, A\rangle : M \to N$ in $\underline{D}^*$, then $\underline{A}_S\langle\alpha, A\rangle =$

$$\begin{array}{ll}
\langle\alpha, S\rangle \cdot A \otimes (\ ) \cdot \langle\alpha, S\rangle^{-1} & \text{for } \underline{A} = \underline{D} \text{ or } \underline{PGL} \\
\alpha \cdot A \otimes (\ ) \cdot \alpha^{-1} & \text{for } \underline{A} = \underline{GL} \text{ or } \underline{ALG}_S E_S \\
\text{identity} & \text{for } \underline{A} = \underline{U}, \underline{PIC} \text{ or } \underline{T}
\end{array}$$

Proposition 1.2 There is a commutative exact sequence of group functors on $\underline{D}^*$

$$\begin{array}{ccccccccc}
 & & 1 & & 1 & & 0 & & \\
 & & \uparrow & & \uparrow & & \uparrow & & \\
1 & \longrightarrow & \underline{PGL}_S & \longrightarrow & \underline{D}_S & \longrightarrow & \underline{T}_S & \longrightarrow & 0 \\
 & & \uparrow & & \uparrow & & \uparrow & & \\
 & & \underline{GL}_S & \longrightarrow & \underline{ALG}_S E_S & \xrightarrow{RZ} & \underline{PIC}_S & & \\
 & & \uparrow & & \uparrow & & \uparrow & & \\
 & & \underline{U}_S & & 1 & & 0 & & \\
 & & \uparrow & & & & & & \\
 & & 0 & & & & & &
\end{array}$$

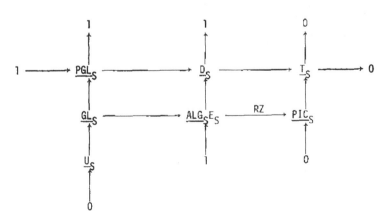

Proof. Set $M = N$ in Theorem 1.1. Naturality and homomorphism properties are easily checked.

§2. Non-abelian Amitsur cohomology sets and sequences

For $0 \le i \le n$ integers define $\varepsilon_i : S^n = S \otimes \cdots \otimes S \to S^{n+1}$ by $\varepsilon_i = S^i \otimes \varepsilon \otimes S^{n-i}$ where $\varepsilon : R \to S$ is the algebra structure map. For $M \in \underline{M}(S^n)$, let $M_i = S^{n+1} \otimes_{S^n} M$ where S^{n+1} is an S^n-algebra via ε_i. Similarly for $f : M \to N$, $f_i = S^{n+1} \otimes_{S^n} f$. The cosemisimplicial identities

$$\varepsilon_i \varepsilon_j = \varepsilon_j \varepsilon_{i-1} \qquad \text{for } i > j$$

yield $M_{ji} \simeq M_{i-1,\, j}$. In particular for $M \in \underline{M}(S)$ we have $M_{01} \simeq M_{00}$, $M_{02} \simeq M_{10}$, $M_{11} \simeq M_{12}$. For $V \in \underline{M}(R) = \underline{M}(S^0)$, let $\lambda_V : V_{01} \to V_{00}$ be the canonical S-isomorphism.

For $\underline{A} \neq \underline{D}^*$ any of the functors in §1 and $M \in \underline{M}(S)$, define the cocycles $Z^1(S/R, \underline{A}_S M) = \{a \in \underline{A}_{S^2}(M_1, M_0) : a_0 a_2 = a_1\}$. Note $a_0 a_2 = a_1$ means the diagram

$$\begin{array}{ccccc} M_{12} & \xrightarrow{a_2} & M_{02} \simeq M_{10} & \xrightarrow{a_0} & M_{00} \\ \Vert & & & & \Vert \\ M_{11} & & \xrightarrow{\qquad a_1 \qquad} & & M_{01} \end{array}$$

is commutative. $\underline{A}_S(M)$ operates on $Z^1(S/R, \underline{A}_S M)$ by $b(a) = b_0 a b_1^{-1}$. The Amitsur cohomology set is

$$H^1(S/R, \underline{A}_S M) = Z^1(S/R, \underline{A}_S M)/\underline{A}_S(M).$$

Then $H^1(S/R, \underline{A}_S M)$ is a functor on $\underline{D}^*$ as follows.
For $\langle\alpha, A\rangle : M \to N$ and $a \in Z^1(S/R, \underline{A}_S M)$ let $\langle\alpha, A\rangle a = \alpha_0 \cdot Aa \cdot \alpha_1^{-1}$ where $Aa = A \otimes a \in \underline{A}_{S^2}(AM_1, AM_0)$.

[Note we are writing AM_1 for $A \otimes M_1 \simeq (A \otimes M)_1$.]

We define H^2 only for $\underline{A} = \underline{U}$ by $Z^2(S/R, \underline{U}_S M) = \{u \in \underline{U}_{S^3}(M_{10}, M_{00}) :$ $u_0 u_2 = u_1 u_3\}$. $\underline{U}_{S^2}(M)$ operates via $v(u) = v_0 v_2 v_1^{-1} u$. Then

$$H^2(S/R, \underline{U}_S M) = Z^2(S/R, \underline{U}_S M)/\underline{U}_{S^2}(M)$$

is a constant (abelian group) functor on $\underline{D}^*$.

In our situation H^0 seems the most difficult cohomology set to define. We shall only define it for $\underline{A} = \underline{PIC}$. For $M \in \underline{M}(S)$, let $H^0(S/R, \underline{PIC}_S M) = \{\langle J, M, V\rangle : J \otimes_S M \simeq V_0, V \in \underline{M}(R), J \in PIC(S)$ and $J_0 \simeq J_1\}$ where $\langle J, M, V\rangle = \langle J', M, V'\rangle$ if $J \simeq J'$ and $V \simeq V'$. For $\langle \alpha, A\rangle : M \to N$ in $\underline{D}^*$, let $\langle \alpha, A\rangle\langle J, M, V\rangle = \langle J, N, AV\rangle$. Then $H^0(S/R, \underline{Pic}_S M)$ is a set functor on $\underline{D}^*$. Note $Pic(R)$ operates on $H^0(S/R, \underline{PIC}_S M)$ by $\langle K\rangle\langle J, M, V\rangle = \langle KJ, M, KV\rangle$.

Proposition 2.1 There are $\underline{D}^*$-natural sequences

$$(2.2) \qquad H^0(S/R, \underline{PIC}_S M) \xrightarrow{\partial^0_M} H^1(S/R, \underline{PGL}_S M) \xrightarrow{\Psi_M} H^1(S/R, \underline{D}_S M) \xrightarrow{\Gamma_M} H^1(S/R, \underline{PIC}_S M)$$

$$(2.3) \qquad H^1(S/R, \underline{GL}_S M) \xrightarrow{\Lambda_M} H^1(S/R, PGL_S M) \xrightarrow{\partial^1_M} H^2(S/R, \underline{U}_S M)$$

∂^0_M depends only on the $Pic(R)$ orbit. When $M = V_0$ for $V \in \underline{M}(R)$, both sequences are exact sequences of pointed sets.

Proof. ∂^0_M is defined by $\partial^0_M \langle J, M, V\rangle = \langle \beta, J\rangle_0^{-1}\lambda_V\langle \beta, J\rangle_1$ where $\beta: J \otimes_S M \to V_0$ is any S-isomorphism. Since $End_S J \simeq S$, any other $\beta' = \beta \cdot J \otimes_S \gamma$ for some $\gamma \in \underline{GL}_S(M)$. Thus $\langle \beta', J\rangle = \langle \beta, J\rangle\langle \gamma, S\rangle$ and so β and β' yield cohomologous cocycles. Thus ∂^M_0 is independent of β. For $\langle K\rangle \in Pic(R)$, the commutative diagram

$$\begin{array}{ccccccc}
M_1 & \xrightarrow{\langle \beta, J\rangle_1} & V_{01} & \xrightarrow{\lambda_V} & V_{00} & \xleftarrow{\langle \beta, J\rangle_0} & M_0 \\
\| & & \downarrow{\scriptstyle \langle 1, K\rangle_{01}} & & \downarrow{\scriptstyle \langle 1, K\rangle_{00}} & & \| \\
M_1 & \xrightarrow{\langle K\beta, KJ\rangle} & (KV)_{01} & \xrightarrow{\lambda_{KV}} & (KV)_{00} & \xleftarrow{\langle K\beta, KJ\rangle_0} & M_0
\end{array}$$

shows $\partial^0_M K\langle J, M, V\rangle = \partial^0_M \langle J, M, V\rangle$.

The definition of the other maps and their naturality in $\underline{D}^*$ is straight forward. The exactnesses of the sequences is also straight forward with some details in [G]. The base point of $H^1(S/R, \underline{A}_S V_0)$ is clearly $\lambda_V : V_{01} \to V_{00}$.

Remark. The base point of $H^i(S/R, \underline{A}_S V_0)$ is <u>not</u> in general preserved by a $\underline{D}^*$-map $V_0 \to W_0$; it is preserved by the image of a $\underline{D}(R/R)$ map $V \to W$.

§3. D* limits and PIC(S) orbits

For $\underline{A}$ any of our functors, let $H^i(\underline{A}_S) = \lim_{\underline{D}^*} H^i(S/R, \underline{A}_S M)$ where $\lim_{\underline{D}^*}$ always means the set-theoretic colimit. Note that for any $P \xleftarrow{<\alpha,\ A>} M \xrightarrow{<\beta,\ B>} Q$ in $\underline{D}^*$, we can construct a commutative square

$$\begin{array}{ccc} M & \longrightarrow & P \\ \downarrow & & \downarrow \\ Q & \longrightarrow & ABM \end{array}$$

Thus $H^i(\underline{A}_S) = \{[a]_M : a \in H^i(S/R, \underline{A}_S M)\}$ where $[a]_M = [b]_N$ iff there are $\underline{D}^*$ maps $M \to P \leftarrow N$ with a and b having the same image in $H^i(S/R, \underline{A}_S P)$.

Proposition 3.1 Each $H^i(A_S)$ is an abelian monoid with $[a]_M \cdot [b]_P = [a \otimes b]_{M \otimes_S P}$ where $<J, M, V> \otimes <L, P, W> = <J \otimes_S L, M \otimes_S P, V \otimes W>$.

Proof. We will only verify the case $i = 1$. The other cases are easier. First note that for $<\alpha, A> : M \to N$, then $<\alpha, A> \otimes P = <\alpha \otimes_S P, A> : M \otimes_S P \to N \otimes_S P$ in $\underline{D}^*$ and for a, b any M, P cocycles respectively, $a \otimes_{S^2} b$ is an $M \otimes_S P$ cocycle and

$$<\alpha, A> a \otimes_{S^2} b = (<\alpha, A> \otimes P)(a \otimes_{S^2} b).$$

Thus the multiplication is well defined on $Z^1(\underline{A}_S) = \lim_{\underline{D}^*} Z^1(S/R, \underline{A}_S M)$. For $c \in \underline{A}_S(M)$,

$$c_0 a c_1^{-1} \otimes b = (c \otimes P)_0 \cdot a \otimes b \cdot (c \otimes P)_1^{-1} \sim a \otimes b$$

and thus the multiplication on $Z^1(\underline{A}_S)$ preserves cohomology classes. Since $<\sigma, R>(a \otimes b) = b \otimes a$ where $\sigma : M \otimes_S P \to P \otimes_S M$ is the "switch map", we see the multiplication is commutative. Finally clearly $[\lambda_R]_S$ is the identity.

Proposition 3.2 There is a natural (in S) equivalence $H^0(\underline{Pic}_S) \to H^0(S/R, Pic)$.

Proof. Let $\theta[J, M, V] = \langle J\rangle$ - clearly a well defined homomorphism $H^0(\underline{PIC}_S) \to H^0(S/R, Pic)$. Its inverse is $\psi\langle J\rangle = [J, J^*, R]$. Clearly $\theta\psi$ is the identity. For any $[J, M, V]$, there is an S-isomorphism $\alpha : V \otimes J^* \to M$. Then $\langle\alpha, V\rangle[J, J^*, R] = [J, M, V]$ and so $\psi\theta$ is also the identity. The maps are clearly natural in S.

Note we use $H^i(S/R, F)$ for ordinary Amitsur cohomology and we let $\bar{H}^0(S/R, F) = H^0(S/R, F)/F(R)$ be the reduced cohomology.

Proposition 3.3 There is a natural equivalence $H^2(\underline{U}_S) \to H^2(S/R, U)$.

Proof. $H^2(S/R, \underline{U}_S M)$ is constantly equal to $H^2(S/R, U)$.

Unfortunately we cannot in general so easily identify the monoids $H^1(\underline{A}_S)$.

For $L \in \underline{PIC}(S)$ and $\underline{A} = \underline{D}$ or $\underline{Pic}$ and $\langle a, J\rangle$ in $\underline{A}_{S^2}(M_1, M_0)$, let $L\langle a, J\rangle = \langle L_0 a, L_0 J L_1^* \rangle$ in $\underline{A}_{S^2}(M_1, M_0)$. The the following facts are easily verified.

Lemma 3.4

1. $L : H^1(S/R, \underline{A}_S M) \to H^1(S/R, \underline{A}_S LM)$ is well defined
2. For $\langle a, J\rangle$ in $\underline{A}_{S^2}(M_1, M_0)$ and $\langle b, K\rangle$ in $\underline{A}_{S^2}(P_1, P_0)$ we have

   $L\langle a, J\rangle \otimes \langle b, K\rangle = L(\langle a, J\rangle \otimes \langle b, K\rangle) = \langle a, J\rangle \otimes L\langle b, K\rangle$
3. For $\Gamma_M : H^1(S/R, \underline{D}_S M) \to H^1(S/R, \underline{Pic}_S M)$ we have $L \cdot \Gamma_M = \Gamma_{LM} \cdot L$

Corollary 3.5 Pic(S) operates on $H^1(\underline{A}_S)$ such that $H^1(\underline{A}_S)/Pic(S)$ is a monoid and Γ_M induces a monoid homomorphism $\Gamma' : H^1(\underline{D}_S)/Pic(S) \to H^1(\underline{PIC}_S)/Pic(S)$.

We can partially identify $H^1(\underline{PIC}_S)$ by

Lemma 3.6 The inclusion $Z^1(S/R, \underline{Pic}_S M) \subseteq Pic(S^2)$ induces a monoid homomorphism $\Gamma'' : H^1(\underline{Pic}_S)/Pic(S) \to H^1(S/R, Pic)$.

We can finally state our main result of this section.

Theorem 3.7 The sequences (2.2) and (2.3) and the maps of (3.2), (3.3), (3.5) and (3.6) induce exact monoid sequences

(3.8) $\overline{H}^0(S/R, \mathrm{Pic}) \xrightarrow{\partial^0} H^1(\underline{PGL}_S) \xrightarrow{\psi} H^1(\underline{D}_S)/\mathrm{Pic}(S) \xrightarrow{\Gamma} H^1(S/R, \mathrm{Pic})$

(3.9) $H^1(\underline{GL}_S) \xrightarrow{\Lambda} H^1(\underline{PGL}_S) \xrightarrow{\partial^2} H^2(S/R, U)$

Proof. We only verify exactness of (3.8). Exactness at $H^1(\underline{PGL}_S)$: For $J \in H^0(S/R, \mathrm{Pic})$, $\psi\partial^0[J] = \psi_{J*}\partial^0{}_{J*}\langle J, J^*, R\rangle = [\langle e, J\rangle_0{}^{-1}\langle e, J\rangle_1]_{J*} = [\langle 1, J_0{}^*J_1\rangle]_{J*} = J^*[\langle\lambda_R, S^2\rangle]_{S^2} = 1$, where $e : J \otimes_S J^* \to S$ is evaluation. Now let $[\langle a, S^2\rangle]_M$ be in $\mathrm{Ker}\,\psi$. Then (after perhaps replacing M by an M' it maps to in $\underline{D}^*$) we can assume there are $\langle\beta, K\rangle$ in $\underline{D}_S(M)$, $J \in \underline{PIC}(S)$ and $\langle\alpha, A\rangle : J \to M$ in $\underline{D}^*$ with $[\langle b, K\rangle_0\langle a, S^2\rangle\langle b, K\rangle_1{}^{-1}]_M = \langle\alpha, A\rangle J[\langle\lambda_R, S^2\rangle]_S$. Then $\langle b, K\rangle_0\langle a, S^2\rangle\langle b, K\rangle_1{}^{-1} = \langle\alpha, J\rangle_0\lambda_A\langle\alpha, J\rangle_1{}^{-1}$ and hence $\langle a, S^2\rangle = \langle\gamma, L\rangle_0{}^{-1}\lambda_A\langle\gamma, L\rangle_1$ where $\langle\gamma, L\rangle = \langle\alpha, J\rangle^{-1}\langle\beta, K\rangle$. Thus $\langle a, S^2\rangle$ is $\partial_M{}^0[L, M, A]$.

Exactness at $H^1(\underline{D}_S)/\mathrm{Pic}(S)$ is easier: $\Gamma\psi[\langle a, S^2\rangle]_M = [S^2] = 1$. For $[\langle a, J\rangle]_M$ in $\mathrm{Ker}\,\Gamma$ we know $[J] \sim 1$ and so $J = L_1L_0{}^*$. Hence $L[\langle a, J\rangle]_M = [\langle L_0a, L_0JL_1{}^*\rangle]_{LM} = \psi[\langle L_0a, S^2\rangle]_{LM}$.

4. Faithfully flat and isotrivial S.

We shall employ Villameyer and Zelinsky's notation and call S isotrivial if there is an S-module which as an R-module is faithfully projective. Examples are 1. those S which themselves are faithfully projective R-modules, 2. polynomial rings and more generally 3. any commutative augmented R-algebra.

Lemma 4.1 $\partial^2 : H^1(\underline{PGL}_S) \to H^2(S/R, U)$ is onto when S is isotrivial.

Proof. For $u \in Z^2(S/R, U)$ and ${}_SV \in \underline{M}(R)$ let $M = V_0$ and let $a = u \otimes_S V : S^3 \otimes_S V \to S^3 \otimes_S V$ where we consider each S^n an S-algebra via the last S-factor. Note $V_{01} \simeq S^2 \otimes V \simeq S^3 \otimes_S V \simeq V_{00}$ and $a_0a_2 = (u_0u_2) \otimes_S V = (u_1u_3) \otimes_S V = u_1 \otimes V \cdot u_3 \otimes V = a_1 \cdot u$. Thus $\partial^2[\langle a, S^2\rangle]_M = [u]$.

We define two "descent" functors on $\underline{D}^*$ as follows. $V_{MOD}(S/R, M)$ is the set of isomorphism class $<V>_M$ of $V \in \underline{M}(R)$ with $V_0 \simeq M$. For $<\alpha, A> : M \to N$ in $\underline{D}^*$, $<\alpha, A><V> = <A \otimes V>$. We can define an abelian monoid structure on $V_{MOD}(S/R) = \lim_{\underline{D}^*} V_{MOD}(S/R, M)$ by $[V]_M \cdot [W]_P = [V \otimes W]_{M \otimes_S P}$ where $[V]_M$ is the the image of $<V>_M$ in $V_{MOD}(S/R)$. The identity element is clearly $[R]_S$.

$V_{ALG}(S/R, E_S M)$ is the set of isomorphism classes $<V>_M$ of faithfully projective R-algebras V with $V_0 \simeq E_S M$. Functoriality is given by $<\alpha, A><V>_M = <V \otimes E_R A>_N$ for $<\alpha, A> : M \to N$ in $\underline{D}^*$. Again $V_{ALG}(S/R) = \lim_{\underline{D}^*} V_{ALG}(S/R, E_S M)$ is an abelian monoid with $1 = [R]_S$.

The first of these monoids is trivial. The second is related to the Brauer group.

Lemma 4.2 $V_{MOD}(S/R) = 1$.

Proof. $[V]_M = [V]_{V_0} = <1, V>[R]_S = 1$.

We have finally come to the Brauer group $B(R)$ - the group of classes $[A]$ of Azumaya R-algebras with $[A] = [B]$ if $A \otimes E_R V \simeq B \otimes E_R W$ for some $V, W \in \underline{M}(R)$. $B(S/R)$ is the subgroup of classes split by S - that is those $[A]$ with $A_0 \simeq E_S M$ for some $M \in \underline{M}(S)$. See [AG] or [KO] for details. We let $Pic_T(S)$ be the torsion subgroup of $Pic(S)$ and set $P_0(S) = Pic(S)/Pic_T(S)$.

Proposition 4.3 There is an exact sequence of abelian groups
$P_0(R) \xrightarrow{\theta_1} P_0(S) \xrightarrow{\theta_2} V_{ALG}(S/R) \xrightarrow{\theta_3} B(S/R) \longrightarrow 0$ where $\theta_3[V]_M = [V]$, $\theta_2[J] = [R]_J$ and $\theta_1[K] = [K_0]$.

Proof. θ_3 is well defined since given $[V]_M$ and $<\alpha, A> : M \to N$, $\theta_3<\alpha, A>[V]_M = \theta_3[V \otimes E_R A]_N = [V \otimes E_R A] = \theta_3[V]_M$. θ_3 is clearly onto and $[V]_M$ is in $\mathrm{Ker}\ \theta_3$ if $V \simeq E_R A$ for some $A \in \underline{M}(R)$. Then since $E_S A_0 \simeq E_S M$ we have an isomorphism $\alpha : A \otimes J \to M$. Then $[V]_M = [E_R A]_M = <\alpha, A>[R]_J = \theta_2[J]$. Finally $\theta_2[J] = [R]_J = 1$ if there is $<\alpha, A> : J \to B_0$ with $A, B \in \underline{M}(R)$ and $<E_R A>_{B_0} = <E_R B>_{B_0}$. Thus (by Theorem 1.1) there is $K \in \underline{PIC}(R)$ with $B \simeq K \otimes A$. Hence $J \otimes A \simeq B_0 \simeq K_0 \otimes A$ and so $JK^* \otimes A \simeq A$. Thus $<JK^*> \in Pic_T(S)$. (See for example [G].) Therefore $[J] = [K_0] = \theta_1[K]$.

We now look at faithfully flat S.

Proposition 4.4 For each $M \in \underline{M}(S)$, there are maps

$\Sigma_M : V_{MOD}(S/R, M) \to H^1(S/R, \underline{GL}_S M)$

$\Sigma'_M : V_{ALG}(S/R, E_S M) \to H^1(S.R, Aut_S E_S M) \simeq H^1(S/R, \underline{D}_S M)$ defined by

$\Sigma_M \langle V \rangle_M = a_0 \lambda_V a_1^{-1}$ for $a : V_0 \xrightarrow{\simeq} M$ and $\Sigma'_M \langle V \rangle_M = a_0 \lambda_V a_1^{-1}$ for $a : V_0 \xrightarrow{\simeq} E_S M$.

They induce homomorphisms $V_{MOD}(S/R) \to H^1(\underline{GL}_S)$ and $V_{ALG}(S/R) \to H^1(\underline{D}_S)$. The latter also preserves the Pic(S) structure. When S is faithfully flat all the above maps are equivalences.

Proof. The facts about Σ_M and Σ'_M are well known in faithfully flat descent theory. The facts on the induced maps easily follow.

Corollary 4.5 $\partial^2 : H^1(\underline{PGL}_S) \to H^2(S/R, U)$ is an isomorphism when S is faithfully flat and isotrivial.

Proof. By Proposition 4.4 and Lemma 4.2 $H^1(\underline{GL}_S) = 0$. Thus ∂^2 is one-one. By Lemma 4.1, ∂^2 is onto.

Corollary 4.6 $B(S/R) \simeq H^1(\underline{D}_S)/Pic(S)$ for S faithfully flat.

Proof. By Propositions 4.3 and 4.4.

We can now prove the Chase Rosenberg Theorem.

Theorem 4.7 For S isotrivial and faithfully flat there is an exact sequence of abelian groups

(4.8)
$$0 \to H^1(S/R, U) \to Pic(R) \to H^0(S/R, Pic)$$
$$\to H^2(S/R, U) \to B(S/R) \to H^1(S/R, Pic)$$

Proof. By Theorem 4.4 (with $M = S$) the first line of (4.8) is exact. By Corollaries 4.5 and 4.6 we can identify the terms of the second line with those of (3.8). It only remains to show $\partial^0 : H^0(\underline{PIC}_S)/\mathrm{Pic}(R) \to H^1(\underline{PGL}_S)$ is one-one.

For $[J, M, V]$ in Ker ∂^0 we may assume (after perhaps replacing M and V) that $M = W_0$ for some $W \in \underline{M}(R)$ and there is an isomorphism $a : J \otimes W \to V_0$ with $\langle a, J\rangle_0^{-1}\lambda_V\langle a, J\rangle_1 = \lambda_W$. Thus there is an isomorphism $b : J_1 \to J_0$ with

$$\begin{array}{ccc} J_1 \otimes W & \xrightarrow{\;a_1\;} & V_{01} \\ \downarrow{\scriptstyle b \otimes W} & & \downarrow{\scriptstyle \lambda_V} \\ J_0 \otimes W & \xrightarrow{\;a_0\;} & V_{00} \end{array}$$

commutative. Thus $b \otimes W$ satisfies the cocycle identity and so since $W \in \underline{M}(R)$ also b is a cocycle. Then by Proposition 4.4, $J = K_0$ where K must certainly be rank one. Thus $[J, M, V] = [K_0, W_0, V] \sim 1$.

§5. Relations with Hoobler's Brauer Group.

Let $\underline{D}_0$ be the image of $\underline{D}(R/R)$ in $\underline{D}^* = \underline{D}(S/R)$. As functors in $\underline{D}_0$ each $H^i(S/R, \underline{A}_S V_0)$ is a pointed set functor. Clearly $\lim_{\underline{D}_0} H^i(S/R, \underline{A}_S M) = H^i_0(\underline{A}_S)$ is also an abelian monoid. Since $\lim_{\underline{D}_0}$ preserves exact sequences of pointed sets (see [G]), sequences (2.2) and (2.3) induce exact monoid sequences

$$H^0_0(\underline{PIC}_S)/\mathrm{Pic}(R) \to H^1_0(\underline{PGL}_S) \to H^1_0(\underline{D}_S) \to H^1_0(\underline{PIC}_S)$$

$$H^1_0(\underline{GL}_S) \to H^1_0(\underline{PGL}_S) \to H^2_0(\underline{U}_S)$$

As before we can identify $H^2_0(\underline{U}_S)$ with $H^2(S/R, U)$ and we show in [G] that $H^i_0(\underline{PIC}_S) \simeq H^i(S/R, \mathrm{Pic}_T)$. Thus we prove in [G] the analogous result to 4.

Theorem 5.1 For S/R isotrivial and faithfully flat, there is an exact sequence

$$0 \to H^1(S/R, U) \to \mathrm{Pic}_T(R) \to H^0(S/R, \mathrm{Pic}_T) \to$$
$$H^2(S/R, U) \to \overline{B}(S/R) \to H^1(S/R, \mathrm{Pic}_T)$$

where Hoobler's Brauer group $\overline{B}(R)$ is the group on Azumaya algebras produced by the equivalence relation $A \sim B$ if $A \otimes M_n \simeq B \otimes M_m$ for matrix rings M_n, M_m.

The two sequences are related as follows

Theorem 5.2 For S/R isotrivial and faithfully flat, there is a commutative exact diagram

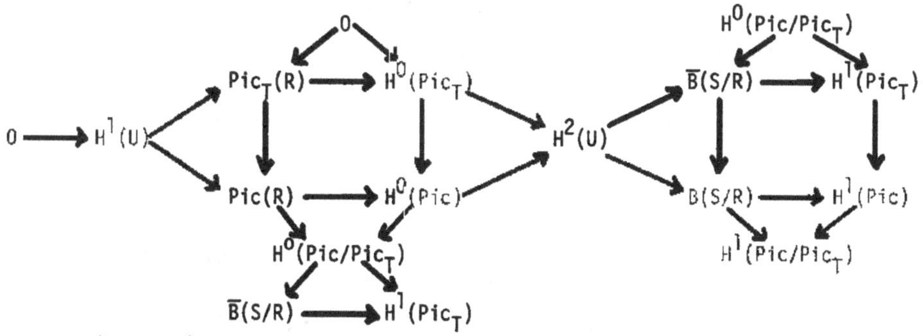

where $H^i(F) = H^i(S/R, F)$.

Proof. Consider the exact K-theory functor sequence $[H_1]$

(5.3) $$0 \to \mathrm{Pic}_T \to \mathrm{Pic} \to \overline{K} \to \overline{B} \to B \to 0$$

where $\overline{K}(S) = K_0(\underline{M}(S))/K_0(\underline{F}(S))$ and $\underline{F}(S)$ is the category of finite rank free modules. Since S has a module $N \in \underline{F}(R)$ the mappings $\phi_n : \overline{K}(S^{n+1}) \to \overline{K}(S^n)$ given by $\phi_n([M]) = [M \otimes_S N]$ is a splitting of the augmented Amitsur complex $C^*(S/R, \overline{K})$. Then the above vertical sequence involving $\overline{B}$ and B is derived from the cohomology of (5.3) and the cohomological triviality of $\overline{K}$. The commutativities are clear.

Remark. The question arises as to whether the different techniques people have used to derive (4.8) yield the same maps. Those of Chase-Rosenberg [CR], Childs [C] and Villamayor-Zelinsky [VZ] are the same since the latter note that spectral sequences and exact couples yield the same sequences. It is easy to check that the directly computed sequences of Auslander-Brumer [AB] and Knus [K] are the same as our (4.8). In the appendix to my thesis $[G_2]$ I showed the Auslander-Brumer and Chase-Rosenberg maps were the same.

REFERENCES

A S. A. Amitsur, Simple Algebras and Cohomology of Arbitrary Fields, Trans. Amer. Math. Soc. 90 (1959), 73-112

AB M. Auslander and A. Brumer, The Brauer group and Galois cohomology of commutative rings, preprint

AG M. Auslander and O. Goldman, The Brauer Group of a Commutative Ring, Trans. Amer. Math. Soc. 97 (1960), 367-409

B H. Bass, Algebraic K-Theory, Benjamin, New York, 1968

CR S. U. Chase and A. Rosenberg, Amitsur Cohomology and the Brauer Group, Mem. Amer. Math. Soc. 52 (1964), 34-78

C L. N. Childs, On normal Azumaya algebras and the Teichmüller cocycle map, J. Algebra 23 (1972), 1-17

G G. S. Garfinkel, A torsion version of the Chase-Rosenberg Exact Sequence, Duke Math. J. 42 (1975), 195-210

G_2 __________, Amitsur cohomology and an exact sequence involving Pic and the Brauer group, Ph.D. Thesis, Cornell University, 1968

H_1 R. T. Hoobler, A Generalization of the Brauer Group and Amitsur Cohomology, Ph.D. Thesis, University of California, Berkeley, 1966

H_2 __________, Cohomology in the Finite Topology and Brauer Groups, Pacific J. Math. 42 (1972), 667-679

K M. Knus, A Teichmüller cocycle for finite extensions, preprint

KO M. Knus and M. Ojanguren, Theorie de la Descente et Algebres d'Azumaya, Springer Lecture Notes 389 (1974)

RZ A. Rosenberg and D. Zelinsky, Automorphisms of Separable Algebras, Pacific J. Math. 11 (1961), 1107-1117

VZ O. E. Villamayor and D. Zelinsky, Brauer groups and Amitsur cohomology for general commutative ring extensions, J. Pure & Applied Algebra, to appear

Long exact sequences and the Brauer group

By

D. Zelinsky

1. The long exact sequences.

In [VZ] the exact sequences of Chase and Rosenberg [CR] of Childs [C] and, in a sense, of Hattori [H] are generalized to a long exact sequence

$$(*)\quad \begin{aligned} &0 \to H^1(S/R,U) \to E_1 \to H^0(S/R,\mathrm{Pic}) \to \cdots \\ &\to H^n(S/R,U) \to E_2 \to H^{n-1}(S/R,\mathrm{Pic}) \to \cdots \end{aligned}$$

together with maps Pic $R \to E_1$ and $\mathrm{Br}(S/R) \to E_2$. When S is faithfully flat over R, then Pic $R \to E_1$ is an isomorphism and $\mathrm{Br}(S/R) \to E_2$ is a monomorphsim; and the latter is an isomorphism if S is faithfully flat over R and "isotrivial": some S-module is finitely generated, faithful and projective over R.

The notation is as follows: $R \to S$ is any extension of commutative rings (with unit). The functors (presheaves) U and Pic from commutative R-algebras to Abelian groups are the usual ones: $U(T)$ is the group of invertible elements of T, and Pic T is the group of isomorphism classes of invertible (projective rank one) T-modules.

For each of the two functors $F = U$ and $F = \mathrm{Pic}$, $H^n(S/R,F)$ denotes Amitsur cohomology, the cohomology of the complex of Abelian groups

$$0 \to F(S) \rightrightarrows F(S \otimes_R S) \mathrel{\substack{\to\\\to\\\to}} F(S \otimes_R S \otimes_R S) \mathrel{\substack{\to\\\to\\\to\\\to}} \cdots$$

The group $\mathrm{Br}(S/R)$ is the relative Brauer group of similarity classes of Azumaya R-algebras which are split by S, that is, the kernel of the map of Brauer groups Br R $\to$ Br S. The groups $E_1, E_2, \ldots$ are explicitly described as refinements of the Amitsur cohomology $H^{n-1}(S/R, \mathrm{Pic})$. Instead of starting with an isomorphism class of invertible S^n-modules (the exponent on S denotes repeated tensor product over R), we begin with a single invertible S^n-module, P. The usual ring homomorphisms ϵ_i: $S^n \to S^{n+1}$ send $x_1 \otimes \cdots \otimes x_n$ to $x_1 \otimes \cdots \otimes x_i \otimes 1 \otimes x_{i+1} \otimes \cdots \otimes x_n$ for $i = 0, \ldots, n+1$, and now associate to P the S^{n+1}-modules $\epsilon_i P = P \otimes_{S^n} S^{n+1}$ with S^{n+1} an S^n-algebra by virtue of the ring homomorphism ϵ_i. Each $\epsilon_i P$ is an invertible S^{n+1}-module and has an inverse $(\epsilon_i P)^{-1}$, namely the dual S^{n+1}-module, $\mathrm{Hom}_{S^{n+1}}(\epsilon_i P, S^{n+1})$. We define the boundary of P, $\delta P = \epsilon_0 P \otimes (\epsilon_1 P)^{-1} \otimes \ldots \otimes (\epsilon_{n+1} P)^{\pm 1}$, with all tensor products over S^{n+1}. Then δ is a functor and $\delta \circ \delta$ is naturally equivalent to the zero functor. That is, for every P there is an S^{n+2}-module isomorphism λ_P: P $\to S^{n+2}$ carrying all isomorphisms between P's into the identity map on S^{n+2}. In the spirit of K-theory, we consider the isomorphism classes of pairs (P, α) with P an invertible S^n-module and α an isomorphism α: P $\to$ identity module, S^{n+1}. We are now in a position to describe the group E_n as a subquotient of this group (under tensor product) of isomorphism classes of pairs: It is the subgroup of isomorphism classes of (P, α) for which $\delta\alpha$: $\delta\delta P \to \delta S^{n+1} = S^{n+2}$ is the natural isomorphism λ_P, reduced modulo the classes of pairs $(\delta Q, \lambda_Q)$ with Q an invertible S^{n-1}-module.

A sketch of the map $\mathrm{Br}(S/R) \to E_2$ is also in order. If A is an R-algebra, denote by ϵA the S-algebra $A \otimes_R S$; we are now using ϵ to denote a functor from R-algebras to S-algebras induced by the ring homomorphism ϵ: R $\to$ S. If A is an Azumaya algebra split by S, then ϵA is Brauer equivalent to S. For Azumaya algebras, Brauer equivalence is the same as Morita equivalence, so we have a Morita equivalence from S to ϵA, that is, an S-equivalence of module categories M: Mod S $\to$ Mod ϵA. The two ring homomorphisms ϵ_0 and ϵ_1

from S to S^2 induce functors everywhere and extend M to two Morita S^2-equivalences $\epsilon_i M$: Mod $S^2 \to$ Mod $\epsilon_i \epsilon A$. But $\epsilon_0 \epsilon A = \epsilon_1 \epsilon A$, so we have an S^2-equivalence $\delta M = \epsilon_1 M^{-1} \circ \epsilon_0 M$ from S^2-modules to S^2-modules. By the Morita theorem, this is just $\cdot \otimes_{S^2} P$ for some invertible S^2-module, P, and we think of δM as being this module P. Proceeding further (but requiring a bit of juggling of natural equivalences), $\delta P = \delta\delta M$ is naturally isomorphic to the identity, so from the splitting of A by S we have produced a pair (P,α), with α an isomorphism of P to S^3. More juggling proves $\delta\alpha = \lambda_P$, so the pair defines an element of E_2. And of course, proof is required that the Brauer class of A determines this element of E_2 uniquely.

In case S is faithfully flat over R, (*) can be derived as a part of an exact couple (replacing a Leray spectral sequence) of a map of sites of R-algebras; the coverings in the first site are just the standard face maps between tensor powers of S as used in constructing Amitsur cohomology; the second site is any one in which all coverings are faithfully flat extensions and which contains enough coverings to split all rank one projective modules (for example, the Zariski coverings). For the present talk, the only relevant item is that the resulting exact couple consists of a sequence of long exact sequences, of which (*) is just the first. The others are also of some interest:

$$(*_q)\qquad \begin{aligned} &0 \to H^1(S/R,C_q) \to H^{q+1}(R,U) \to H^0(S/R,H^{q+1}(\cdot,U)) \to \\ &H^2(S/R,C_q) \to H^1(S/R,C_{q+1}) \to H^1(S/R,H^{q+1}(\cdot,U)) \to \\ &\ldots\ H^i(S/R,C_q) \to H^{i-1}(S/R,C_{q+1}) \to H^{i-1}(S/R,H^{q+1}(\cdot,U)) \to \end{aligned}$$

where C_q is the qth kernel in an injective resolution of the sheaf U, so $C_0 = U$. The cohomology $H^n(\cdot,U)$ is the "ordinary" derived functor cohomology of the sheaf U on the larger site. By the hypotheses on this site (faithful flatness and splitting of rank one projectives) $H^1(\cdot,U) = \text{Pic}$, and [VZ] proves $H^{n-1}(S/R,C_1) = E_n$. With these identifications, $(*_0)$ becomes $(*)$.

However, we get extra information. From $(*_1)$, $0 \to E_2 \to H^2(R,U) \to H^0(S/R,H^2(\cdot,U)) \subset H^2(S,U)$, so

$$E_2 \cong \mathrm{Ker}(H^2(R,U) \to H^2(S,U))$$

with the cohomology computed in any of a large collection of sites. Thus, even if $\mathrm{Br}(R) \to H^2(R,U)$ is not an isomorphism, we have proved $\mathrm{Ker}(\mathrm{Br}(R) \to \mathrm{Br}(S))$ is isomorphic to $\mathrm{Ker}(H^2(R,U) \to H^2(S,U))$ if S is faithfully flat and isotrivial over R.

The groups E_n with $n > 2$ must also be of some use, though I have no answer yet. If Ker_n denotes $\mathrm{Ker}(H^n(R,U) \to H^n(S,U))$, then there is a homomorphism $E_n \to \mathrm{Ker}_n$, because E_n can be identified with $H^{n-1}(S/R,C_1)$, as we just mentioned, and $(*_i)$ includes a homomorphism $H^{n-i-1}(S/R,C_i) \to H^{n-i-1}(S/R,C_{i+1})$, and $(*_{n-1})$ begins with an isomorphism $H^1(S/R,C_{n-1}) \to \mathrm{Ker}_n$. The composite of all these is a map $E_n \to \mathrm{Ker}_n$.

This map is not likely to be an isomorphism. For example, if $n = 3$, $E_3 \to \mathrm{Ker}_3$ is the map obtained by identifying E_3 with $H^2(S/R,C_1)$ and Ker_3 with $H^1(S/R,C_2)$ in $(*_1)$, getting an exact sequence

$$(*_1)' \quad 0 \to H^2(R,U) \to H^0(S/R,H^2(\cdot,U)) \to E_3 \to \mathrm{Ker}_3 \to H^1(S/R,H^2(\cdot,U)).$$

The middle map in this sequence, the one describing the kernel of $E_3 \to \mathrm{Ker}_3$ is a direct generalization of the Teichmüller cocycle.

Recall that Teichmüller's cocycle is a homomorphism defined when S is a Galois field extension of R with group G; it maps $\mathrm{Br}(S)^G$ into $H^3(G,U(S))$; replacing Galois cohomology by Amitsur cohomology leads one to expect a map from $H^0(S/R,\mathrm{Br})$ to $H^3(S/R,U)$. In fact the map goes into E_3, which is isomorphic to $H^3(S/R,U)$ and to $H^3(G,U(S))$ in Teichmüller's situation. (These arguments follow in the path of Childs [C] and Knus [K].) In Teichmüller's case, and in Knus's,

$Br = H^2(\cdot,U)$ at least for the rings R,S and $S \otimes S$, which are the only rings involved in $H^0(S/R,Br)$. Hence our map $H^0(S/R,H^2(\cdot,U)) \to E_3$ is a (possibly the correct) generalization of Teichmüller's. In any case, the image of this map is the kernel of $E_3 \to Ker_3$. The cokernel of the latter is given by the rest of $(*_1)'$.

2. Limits

There are two purposes in the homomorphism $Br(S/R) \to E_2$ and the conditions under which it is a monomorphism or an isomorphism. The resulting description of Br(S/R) allows computation of this kernel $Ker(Br(R) \to Br(S))$, and hence asserts that $Br(R) \to Br(S)$ is a monomorphism or is not a monomorphism in specific cases. If we are interested only in finding monomorphisms $Br(R) \to Br(S)$, then the isotrivial hypothesis is not especially interesting; the map of Brauer groups is a monomorphism if S is faithfully flat over R and if the corresponding E_2 is zero. However, if S is also isotrivial over R, the map of Brauer groups will not be a monomorphism if E_2 is not zero. For these purposes, it is useful to know as many extensions S of R as possible which are faithfully flat and isotrivial.

A second purpose is to use Br(S/R) as a step in computing the full Brauer group Br(R). For this purpose, we are interested in the question: Is Br(R) the union of Br(S/R) for a given collection of S's? The answer is yes for faithfully flat S, so that every Br(R) is the union of certain subgroups of some E_2's. It would be interesting to know what condition on R would guarantee that Br(R) is the union of Br(S/R) with S ranging over the faithfully flat, isotrivial extensions. For such an R, Br(R) would be the union of the corresponding E_2's. In other words, for what R is every Azumaya R-algebra split by a faithfully flat, isotrivial S? Since finitely generated, faithful and projective imply faithfully flat and isotrivial, these R's include all semilocal rings. I do not know, however, whether every algebra that can be split by a faithfully flat isotrivial S can also be split by a finitely generated, faithful, projective S. In other words, if A is the collection of all faithfully flat, isotrivial extensions of R and B is the collection of all extensions which are finitely generated, faithful, projective R-modules, is it true that $\cup_{S \in A} Br(S/R) = \cup_{S \in B} Br(S/R)$? (Since $B \subset A$, the first union contains the second.) This can conceivably be true for Noetherian R, for example, because here is another version of isotriviality:

A ring extension S over R is isotrivial if and only if there is an R-algebra homomorphism from S to a split Azumaya R-algebra (= $End_R(V)$ with V finitely generated, faithful and projective; the resulting S-module structure on V gives the required isotriviality). Now if R is Noetherian, and S maps to a commutative algebra S' contained in $End_R(V)$, then S' is a finitely generated R-module because $End_R(V)$ is, and any algebra split by S is also split by S'. The trouble is that S' need not be projective, nor even contained in any projective, commutative subalgebra of $End_R(V)$. However, it is conceivable that moving to some other V may produce such an S'.

This consideration of the union of Br(S/R) as S ranges over a collection of R-algebras is mirrored by the homological analog, the direct limit $\lim_S H^n(S/R,U)$ taken over the category of covers S of R in the given topology. Here, lim means the direct limit of the sets $H^n(S/R,U)$ as S varies. Since two ring algebra homomorphisms $S \to T$ produce the same group homomorphism $H^n(S/R,U) \to H^n(T/R,U)$, this limit is an Abelian group; group homomorphisms like $H^{n-2}(S/R,Pic) \to H^n(S/R,U)$ in (*) induce group homomorphisms $\lim H^{n-2}(S/R,Pic) \to \lim H^n(S/R,U)$, and the limit of an exact sequence of Amitsur cohomology groups is an exact sequence [A1, Ch. 1]. (Since Amitsur cohomology is the Cech cohomology of a covering, these limits are the full Cech cohomology of the topology.)

As we remarked before, one of the principal results of [VZ] is that E_n is also an Amitsur cohomology group when the covers in the topology are all faithfully flat and can split every rank one projective, namely $E_n = H^{-1}(S/R, C_1)$. Hence the same remarks apply to all the terms in (*), and the direct limit of that exact sequence is an exact sequence connecting the limit groups.

If the topology is the étale topology, then $\lim H^n(\cdot/R, \mathrm{Pic})$ is zero because Pic is a derived functor and because of results of [A2]. The resulting limit of (*) is then simply a collection of bijections $\lim H^{n+1}(\cdot/R, U) \to \lim E_n$. Since in this topology, if the ring R is Noetherian, the Cech cohomology equals the derived functor cohomology $H^{n+1}(R, U)$, we have another description of the latter cohomology group, namely $\lim E_n$, and the limit of our monomorphisms $\mathrm{Br}(S/R) \to E_2$ is the familiar monomorphism $\mathrm{Br}(R) \to H^2(R, U)$.

One should try limits over other topologies. For example, the collections A and B of the previous section have the properties required for good direct limits of Amitsur cohomology, namely closure under $\otimes_R$, which implies axioms L1 and L3 of [A1, Ch. 1]. We get

$\lim_{S \in A} E_2$ = the subgroup of Br(R) consisting of algebra classes with a faithfully flat, isotrivial splitting ring.

$\lim_{S \in B} E_2$ = the subgroup of Br(R) consisting of algebra classes with a finite, faithfully projective splitting ring.

As we mentioned in the preceding section, it would be interesting to know for which rings R, these subgroups of Br(R) exhaust Br(R).

References

[A1] M. Artin, Grothendieck Topologies, Harvard U. Notes, 1962.

[A2] M. Artin, On the joins of Hensel rings, Advances in Math. 7(1971) 282-296.

[CR] S. U. Chase and A. Rosenberg, Amitsur cohomology and the Brauer group, Mem. Amer. Math. Soc. 52(1965) 34-68.

[C] L. N. Childs, On normal Azumaya algebras and the Teichmüller cocycle map. J. Alg. 23(1972) 1-17.

[H] A. Hattori, Certain cohomology associated with Galois extensions of commutative rings, Sci. Papers Coll. Gen. Educ., U. Tokyo 24(1974) 79-91.

[K] M.-A. Knus, On the Teichmüller cocycle, unpublished.

[VZ] O. E. Villamayor and D. Zelinsky, Brauer groups and Amitsur cohomology for general commutative ring extensions, to appear in J. Pure and Applied Algebra.

THE PICARD SEQUENCE OF A FIBRATION*

by

Andy R. Magid

Fix an algebraically closed field k. Pre-varieties over k are to be irreducible. A morphism $f : E \to B$ of pre-varieties over k is a fibration with fiber F if for every $b \in B$ there is an open subset U of B containing b and an isomorphism $F \times U \to f^{-1}(U)$ such that

$$\begin{array}{ccc} F \times U & \xrightarrow{\sim} & f^{-1}U \\ & {\scriptstyle pr_2} \searrow \quad \swarrow {\scriptstyle f} & \\ & U & \end{array}$$

commutes. We call E the total space and B the base space of the fibration. The purpose of this paper is to present an exact sequence relating the Picard groups of the base space, total space, and fibre of a fibration.

---

* From a lecture presented to the Conference of Brauer Groups, Northwestern University, October 13-17, 1975.

The sequence is inspired by a sequence due to R. Fossum and B. Iverson, which relates the relevant Picard group and values of the "relative units functor", U_k, defined by $U_k(w) = \Gamma(w)^*/k^*$ for pre-varieties w. Precisely, Fossum and Iverson show [3, PROP 2.3, p.273] that if $f: E \to B$ is a fibration with fibre F, where E, B, F are smooth varieties and F is rational, Then there is an exact sequence

$$(*) \quad 1 \to U_k(B) \to U_k(E) \to U_k(F) \to \mathrm{Pic}(B) \to \mathrm{Pic}(E) \to \mathrm{Pic}(F) \to 1.$$

It turns out that this sequence is a special case of the following theorem:

<u>Theorem</u>: Let $f : E \to B$ be a fibration with fibre F. Suppose that E, B, F are irreducible, normal prevarities, and that for all sufficiently small open sets w of B the natural map $\mathrm{Pic}(F) \times \mathrm{Pic}(w) \to \mathrm{Pic}(F \times w)$ is an isomorphism. Then there is an exact sequence

$$(**) \quad 1 \to U_k(B) \to U_k(E) \to U_k(F) \to \mathrm{Pic}(B) \to \mathrm{Pic}(E) \to \mathrm{Pic}(F)$$

$$\xrightarrow{\delta} H^2(B, G_m) \xrightarrow{\varepsilon} H^2(E, G_m).$$ (Cohomology is in the Zariski topology).

The proof of the theorem is relatively brief and appears in [6, Thm. 5]. (The arguments in the reference are given for varieties, but actually apply to prevarieties). Here, we will give some applications of the theorem.

To give applications of the theorem, it is necessary to know when the hypothesis $\mathrm{Pic}(F) \times \mathrm{Pic}(W) = \mathrm{Pic}(F \times W)$ is satisfied. The following theorem of Ischebeck is useful here:

Theorem [S, Satz 1.7, p. 143]. Let X and Y be normal varieties. Then there is an exact sequence

$$(***) \quad 1 \to \mathrm{Pic}(X) \times \mathrm{Pic}(Y) \to \mathrm{Pic}(X \times Y) \to \mathrm{Pic}(k(X) \otimes_k k(Y)).$$

A proof of Ischebeck's theorem is given in an appendix below. We observe a corollary (also due to Ischebeck) of the above theorem:

Corollary Let X, Y be normal varieties with X rational. Then the natural map $\mathrm{Pic}(X) \times \mathrm{Pic}(Y) \to \mathrm{Pic}(X \times Y)$ is an isomorphism.

Proof: By the above theorem, it will suffice to show that $\mathrm{Pic}(k(X) \otimes_k k(Y)) = 1$. Suppose $k(X) = k(t_1,\ldots,t_n)$ with $\{t_i\}$ algebraically independent over k. Then $k(X) \otimes_k k(Y)$ is a localization of the regular factorial ring $k(Y)[t_1,\ldots,t_n]$, so there is a surjection $1 = \mathrm{Pic}(k(Y))[t_1,\ldots,t_n] \to \mathrm{Pic}(k(X) \otimes_k k(Y))$.

Because of the corollary, the hypothesis for the sequence (**) will always be satisfied for fibrations with rational fibre.

In particular, if F is rational the sequence (**) gives the Fossum-Iverson sequence (*), except for the right hand term. But if B and F are, in eddition, smooth, the map ε of (**) is an injection [6, Remark] and hence the map δ of (**) has image 1, so (**) reduces to (*) in this case.

We now give a series of examples of the sequence (**).

Example 1 Let B be a normal variety and $E \to B$ a vector bundle of rank n over B. Then $E \to B$ is a fibration which fibre $k^{(n)}$. Now $k^{(n)}$ is rational, and $U_k(k^{(n)}) = \mathrm{Pic}(k^{(n)}) = 1$, so the sequence (**) reduces to, in part,

$$1 \to \mathrm{Pic}(B) \to \mathrm{Pic}(E) \to 1.$$

Thus $\mathrm{Pic}(B) = \mathrm{Pic}(E)$. Suppose further that B is affine, and let $R = k[V]$. Then there is a projective R-module P of rank n such that $k[E] = S_R(P)$ (where the latter denotes the symetric algebra over R of P). Conversely, rank n projective modules over R give rise to vector bundles of rank n over B. Thus we get an affine form of example 1: If R is an affine normal domain over k and P is a projective R module of finite rank, $\mathrm{Pic}(S_R(P)) = \mathrm{Pic}(R)$.

Example 2 Let B be a normal variety and L a line bundle over B. Let $E \to B$ be the associated principal bundle of L over B. Then $E \to B$ is a fibration with fibre k^*, which is rational, so the sequence (**) obtains. Since $U_k(k^*) = \mathbb{Z}$ and $\mathrm{Pic}(k^*) = 1$, (**) reduces to, in part,

$$1 \to U_k(B) \to U_k(E) \to \mathbb{Z} \to \mathrm{Pic}(B) \to \mathrm{Pic}(E) \to 1.$$

Since $E\times_B E$ has a section over E, $[L]$ is in the kernel of $\mathrm{Pic}(B) \to \mathrm{Pic}(E)$, and in fact $[L]$ is the image in $\mathrm{Pic}(B)$ of the generator of $\mathbb{Z}$. Thus $[L]$ will be a torsion class in $\mathrm{Pic}(B)$ exactly when the map $U_k(E) \to \mathbb{Z}$ is non-trivial; i.e., when there is a relative unit on E not coming from B. We examine the significance of this observation when B is affine. If $R = k[B]$, L comes from a rank 1 projective R - module I, and $S = k[E] = \Sigma\, I^{\otimes n}$, the sum extending over all (positive and negative) integers n. If $[L]$ is torsion in $\mathrm{Pic}(B)$, $[I]$ is torsion in $\mathrm{Pic}(R)$, so there is an integer $m > 0$ and an isomorphism $a : I^{\otimes m} \to R$. Let e denote the generator of R^* such that $e(1) = 1$. Choose x in $I^{\otimes m}$ such that $a(x) = 1$ and choose f in $I^{\otimes -m}$ such that $a^*(e) = f$. Then, in S, $f \cdot x = f(x)$ $=(a^*(e))(x) = e(a(x)) = e(1) = 1$, so x is the new relative unit in S which must be present, by the above, when $[L]$ is torsion in $\mathrm{Pic}(B)$.

<u>Example 3</u> Let B be a normal variety and $E \to B$ a locally trivial $\mathbb{P}^{(n)}$ bundle. Then $E \to B$ is a fibration whose fibre $\mathbb{P}^{(n)}$ is rational, so that the sequence (**) applies. Since $U_k(\mathbb{P}^{(n)}) = 1$ and $\mathrm{Pic}(\mathbb{P}^{(n)}) = \mathbb{Z}$, the sequence (**) gives a sequence

$$1 \to \mathrm{Pic}(B) \to \mathrm{Pic}(E) \to \mathbb{Z} \to H^2(B,G_m) \to H^2(E,G_m).$$

In the terminology of [4, p.64], E is (locally trivial) Severi-Brauer fibration, and hence corresponds to an Azumaya algebra $\mathfrak{A}$ over B, locally trivial in the Zariski topology. We can regard $H^2(B, G_m)$ as a subgroup of $H^2(B_{et}, G_m)$ (the cohomological Brauer group of B), and this latter contains the Brauer group of B. According to [4, p. 126], it is possible to show that the class of $\mathfrak{A}$ in the Brauer group of B, regarded as a subgroup of $H^2(B_{et}, G_m)$, is $\delta(1)$. Thus we have the following criterion: $[\mathfrak{A}]$ is trivial in the Brauer group of B if and only if there is a line bundle L on E which restricts to the generator $\sigma_{pn}(1)$ of $\mathrm{Pic}(R^{(n)}) = \mathbb{Z}$. (This result is reported in [4, p.69], presumably obtained by a direct argument).

Example 4. Let $E = k^{(4)} - (k^{(2)} \times (0,0)) - ((0,0) \times k^{(2)})$, with coordinates x_1, x_2, y_1, y_2 and let $B = \{(a_1, a_2, a_3, a_4) \mid a_1 a_4 = a_2 a_3\} - (0,0,0,0)$. ($B$ with the origin is the cone over the Segre embedding $\mathbb{P}^1 \times \mathbb{P}^1 \to \mathbb{P}^3$). Let $f\colon E \to B$ send (x_1, x_2, y_1, y_2) to $(x_1y_1, x_1y_2, x_2y_1, x_2y_2)$. It is easy to check that f is onto the normal variety B. We are going to verify that f is a fibration with fibre k^*. For $i = 1,2,3,4$, let $B_i = B - (a_i = 0)$, so $B = B_1 \cup B_2 \cup B_3 \cup B_4$. Since the property of being a fibration is local on B, we can restrict attention to the open subsets B_i, and since the arguments are the same for each i we can even assume $i = 1$.

Let u_1, u_2, u_3 be coordinates on $k^{(3)}$. Then there is an isomorphism $B_1 \to k^{(3)} - (u_2 = 0)$ given by $(a_1, a_2, a_3, a_4) \to ({}^{a_3}/a_1,\ a_1,\ a_2)$; the inverse isomorphism sends (u_1, u_2, u_3) to $(u_2, u_3, u_1u_2, u_1u_3)$. (To check that the composites both ways are the identity, recall that $a_2({}^{a_3}/a_1) = a_4$ on B_1). Now let $E_1 = f^{-1}(B_1)$; $E_1 = E - (x_1 = 0) - (y_1 = 0)$. There is an isomorphism $E_1 \to k^* \times (k^{(3)} - (u_2 = 0))$ given by sending (x_1, y_1, x_2, y_2) to $(x_1, x_2/x_1, x_1y_1, x_1y_2)$; the inverse isomorphism is given by sending (t, u_1, u_2, u_3) to $(t, tu_1, t^{-1}u_2, t^{-1}u_3)$. Finally, it is easy to check that the following diagram commutes:

$$\begin{array}{ccc} k^* \times (k^{(3)} - (y_2 = 0)) & \to & E_1 \\ {\scriptstyle pr_2}\downarrow & & \downarrow{\scriptstyle f} \\ k^{(3)} - (y_2 = 0) & \to & B_1 \end{array} .$$

Thus $f : E \to B$ is a fibration with fibre k^*, which is rational. As in example 2, this means we have an exact sequence

$$1 \to \mathbb{Z} \to \mathrm{Pic}(B) \to \mathrm{Pic}(E) \to 1.$$

But $\mathrm{Pic}(E) = 1$ (E is open in $k^{(4)}$), so $\mathrm{Pic}(B) = \mathbb{Z}$. Now B is an open subset of the affine variety with coordinate ring $R = k[a_1, a_2, a_3, a_4]/(a_1a_4 - a_2a_3)$, whose complement is the single point $(0,0,0,0)$ which has codimension 3 in B, so $k[B] = R$ and there is an isomorphism

$\mathrm{Pic}(B) \to C\ell(R)$ (the divisor class group). So $C\ell(R) = \mathbb{Z}$. This result is obtained by direct means in [2, PROP 11.4, p. 51].

Example 5 To construct this example, we make a change of coordinates in the description of the B in example 4. B was essentially defined by the relation $a_1a_4 - a_2a_3$ in $k^{(4)}$. Let $a_4^1 = \dfrac{a_1 + a_4}{2}$, $a_2^1 = \dfrac{a_1 - a_4}{2}$, $a_1^1 = {}^1/_2a_2$ and $a_3^1 = a_3(\frac{1}{2} \in k)$, then $a_1a_4 - a_2a_3$ becomes $a_4^{1^2} - a_2^{1^2} + 2a_1^1a_3^1$. Let V be the variety in $k^{(4)}$ given by $V = \{(a_1,a_2,a_3,a_4) \mid a_4^2 = a_2^2 - 2a_1a_3\} - (k \times (0,0,0))$. Then V is essentially an open subset of B whose complement has codimension 2, so $\mathrm{Pic}(V) = \mathrm{Pic}(B) = \mathbb{Z}$. We are going to construct a fibration with total space V and fibre k.

Let $Z = k^{(3)} - (k \times (0,0))$ with coordinates (t,u,v). Let $f : Z \to V$ send (t,u,v) to $(ut + v(t^2/2), u + vt, v, u)$. It is easy to see that f is well-defined, and not difficult to check that f is an open immersion. Now define $\sigma : V \to V$ by $\sigma(a_1,a_2,a_3,a_4) = (a_1,a_2,a_3, - a_4)$. Then σ is an automorphism of V of period 2. We let $V_1 = f(Z)$ and $V_2 = \sigma f(Z)$. Then V_1 is open in V and it is easy to verify that $V = V_1 \cup V_2$. We want to see how V_1 is attached to V_2 in V. Now $x \in V_1 \cap V_2$ if and only if there are t,u,v,t^1,u^1,v^1 such that $x = f(t,u,v) = \sigma\ (t^1,u^1,v^1)$, and it is easy to check that this happens if and only if $v \neq 0$,

$u^1 = -u$, and $t^1 = t + 2(u/v)$. Now V_i looks like 3-space with the t-axis deleted, $V_1 \cap V_2$ in V_1 looks like V_1 with the (t,u) plane deleted, and, in particular, points in the (u,v) plane of V_1 off the u axis, are attached to points in V_2 via $(0,u,v) \to (2u/v, -u, v)$. (See figure 1). Let w_1 (resp w_2) be planes with the origin deleted and let $h_i : V_i \to w_i$ be the projection onto the (u,v) plane $(i = 1)$ and (u^1,v^1) plane $(i = 2)$. Attach points in w_1 off the u axis to points in w_2 by reflection about the v-axis : $(u,v) \to (-u,v)$ (see figure 2). The resulting prevariety w looks like a plane with the origin deleted and the u axis doubled (figure 3). Since the identification producing w out of w_1 and w_2 is compatible, under the h_i, with the identification producing V out of V_1 and V_2, h_1 and h_2 fit together to give a morphism $h : V \to w$ whose restriction to w_i gives $h_i : V_i \to w_i$. Now V_i looks like $k \times w_i$ and this isomorphism is compatible with h_i, so h is a fibration with fibre k. Of course k is rational, so we can use the sequence (**). Since $U_k(k) = \mathrm{Pic}(k) = 1$, (**) reduces to, in part

$$1 \to \mathrm{Pic}(w) \to \mathrm{Pic}(V) \to 1.$$

We already observed that $\mathrm{Pic}(V) = \mathbb{Z}$, hence $\mathrm{Pic}(w) = \mathbb{Z}$. Now w is non-singular (even though it is not separated) since it is covered by the non-singular open sets w_1 and w_2 which are planes minus the origin, so the divisor class group of w

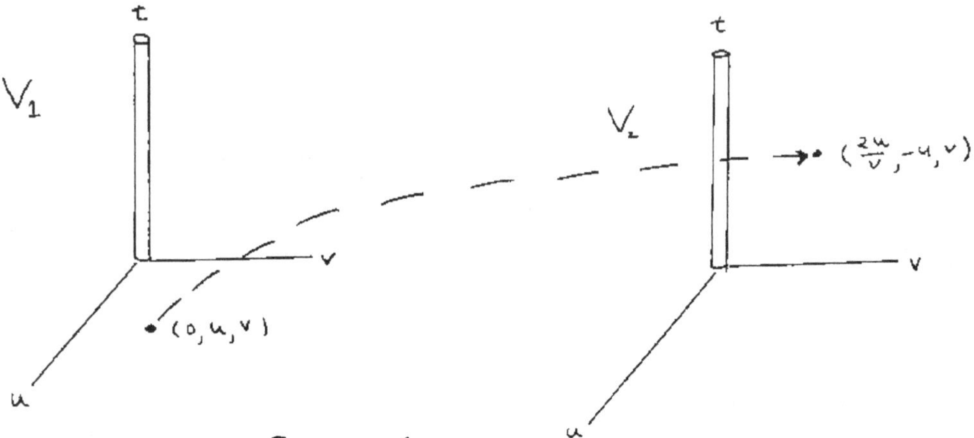

Figure 1

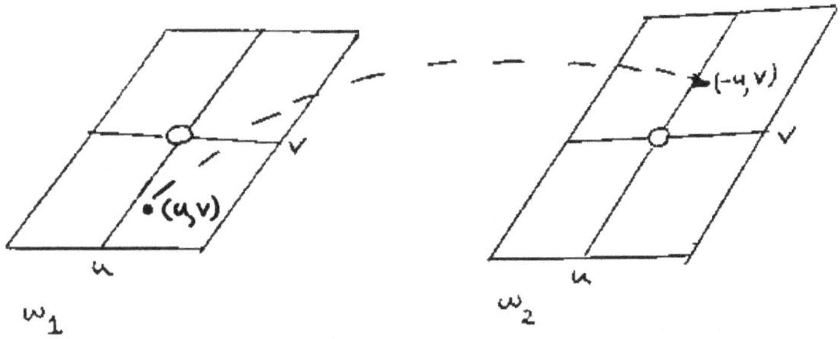

Figure 2

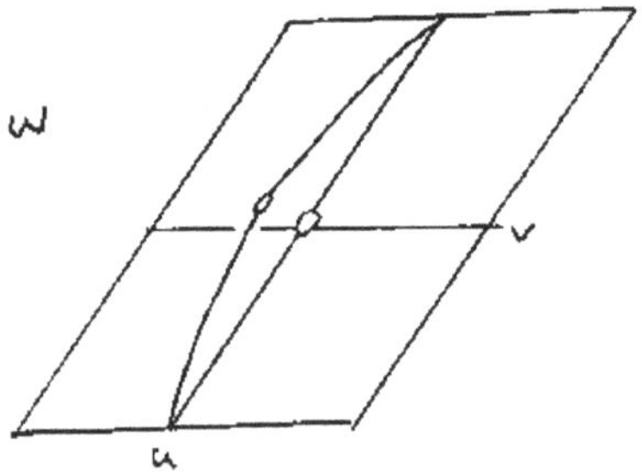

Figure 3

is the same as Pic(w). Since the latter is $\mathbb{Z}$, there is a divisor on w which is not the divisor of a function. It is easy to see this divisor (figure 3) : it is the extra copy of the u axis. The candidate for a function which has the extra u axis as its divisor is v, but the zeros of v are both copies of the u axis, and there is no function separating the copies.

(Example 5 is joint work of the author and Amassa Fauntleroy. Fauntleroy discovered the variety V in the course of a study of actions of the additive group G_a of k , regarded as an algebraic group over k, on quasi-affine (=open in an affine) varieties over k. He observed that G_a acts on the quasi-affine variety V such that in each V_i, G_a acts by translation in the t coordinate, and that w is the geometric quotient of V by G_a. Thus G_a can act locally trivially on a quasi-affine variety such that the quotient is not separated, and hence not quasi-projective. Fauntleroy has further used this example to give an example of an action of G_a on a quasi-affine variety which has no fixed points but such that the geometric quotient fails to exist, even as a prevariety.)

<u>Appendix</u>: A proof of Ischebeck's Theorem.

In this appendix we present a cohomological proof of Ischebeck's exact sequence (***).

We begin by recalling the following special case of [3,PROP 1.1, p.271]:

Lemma 1 Let R and S be Krull domains over k such that $R \otimes_k S$ is a Krull domain, and let E be the quotient field of R. Then there is an exact sequence of divisor class group

$$C\ell(R) \to C\ell(R \otimes_k S) \to C\ell(E \otimes_k S).$$

This result has the following corollary:

Lemma 2 Let R and S be affine normal domains over k with quotient fields K and L respectively, and let M and N be maximal ideals of R and S. Then the natural map $\mathrm{Pic}(R_M \otimes_k S_N) \to \mathrm{Pic}(K \otimes_k L)$ is a monomorphism.

Proof: $R_M \otimes S_N = [(R - M) \otimes (S - N)]^{-1}(R \otimes S)$ is a Krull domain, and hence there is an exact sequence $C\ell(R_M) \to C\ell(R_M \otimes S_N) \to C\ell(K \otimes S_N)$ by lemma 1. Let $[I]$ be in the kernel of $\mathrm{Pic}(R_M \otimes S_N) \to \mathrm{Pic}(K \otimes S_N)$. The exact sequence shows that I is induced from a divisorial ideal of R_M. But $R_M \otimes S_N$ is faithfully flat over R_M, so this divisorial ideal is invertible, since I is. But invertible ideals of R_M are principal, hence $[I] = 1$. Now $K \otimes S_N = [(R - 0) \otimes (S - N)]^{-1}(R \otimes S)$ is a Krull domain, so there is by lemma 1 an exact sequence $C\ell(S_N) \to C\ell(K \otimes S_N) \to C\ell(K \otimes L)$. Exactly as above, this implies that $\mathrm{Pic}(K \otimes S_N) \to \mathrm{Pic}(K \otimes L)$ is injective. The composite of the two injections is the natural map of the statement of the lemma, and hence the result follows.

Theorem 3 Let X and Y be normal k pre-varieties. Then there is an exact sequence

$$1 \to \mathrm{Pic}(X) \times \mathrm{Pic}(Y) \to \mathrm{Pic}(X \times Y) \to G \to H^2(X,U_k) \times H^2(Y,U_k)$$

$$\to H^2(X \times Y,U_k);$$

moreover, the group G is a subgroup of $\mathrm{Pic}(k(X) \otimes_k k(Y))$.

Proof: Let $\mathcal{U}$ denote the open cover of $X \times Y$ by all open sets of the form $U \times V$ where U is an open affine in X and V is an open affine Y. The Cech-to-derive functor cohomology spectral sequence [1, 3.1, p. 37] for the Zariski sheaf U_k is $H^p(\mathcal{U},H^q(\_\ , U_k)) \Longrightarrow H^n(X \times Y,U_k)$. The exact sequence of low degree terms gives

$$1 \to H^1(\mathcal{U},U_k) \to H^1(X \times Y,U_k) \to H^0(\mathcal{U}, H^1(\_\ , U_k)) \to H^2(\mathcal{U},U_k)$$

$$\to H^2(X \times Y,U_k).$$

For $U \times V$ in $\mathcal{U}$, Rosenlicht's lemma [3, Lemma 2.1, p.272] implies that $U_k(U \times V) = U_k(U) \times U_k(V)$. It follows that for any i the Cech cochain group $C^i(\mathcal{U},U_k)$ is $C^i(X,U_k) \times C^i(Y,U_k)$, and hence that $H^i(\mathcal{U},U_k) = H^i(X,U_k) \times H^i(Y,U_k)$. Since $H^1(X,U_k) = H^1(X,U_k) = \mathrm{Pic}(X)$, and similarly for Y, $H^1(\mathcal{U},U_k) = \mathrm{Pic}(X) \times \mathrm{Pic}(Y)$. Also, $\mathrm{Pic}(X \times Y) = H^1(X \times Y,U_k)$. (We are using [6, Lemma 2(iii)] in these assertions.) The group G of the theorem is $H^0(\mathcal{U},H^1(\_\ , U_k))$. To complete the proof, we must embed G in $\mathrm{Pic}(k(X) \otimes_k k(Y))$.

For $U \times V$ in $\mathcal{U}$, $H^1(\_, U_k)(U \times V) = H^1(U \times V, U_k)$ $= \mathrm{Pic}(U \times V)$, and restriction to the generic points of X and Y induces a map $\mathrm{Pic}(U \times V) \to \mathrm{Pic}(k(X) \otimes k(Y))$. Thus there is a map from $H^1(\_, U_k)$ to the constant sheaf on $X \times Y$ with value $\mathrm{Pic}(k(X) \otimes k(Y))$. To embed G in the latter we show that H^0 of this map is injective. It is enough to check this at stalks. For (x,y) in $X \times Y$ the $\mathcal{U}$ - stalk of $H^1(\_, U_k)$ is $\mathrm{Pic}(\mathcal{O}_{X,x} \otimes \mathcal{O}_{Y,y})$, and lemma 2 shows that this groups maps injectively to $\mathrm{Pic}(k(x) \otimes_k k(Y))$. Now we apply $H^0(\mathcal{U}, \_)$ to obtain the embedding of G.

REFERENCES

1. M. Artin, Grothendieck Topologies, Harvard University Department of Mathematics Lecture Notes, 1962.

2. R. Fossum, The Divisor Class Group of a Krull Domain, Springer-Verlag, New York, 1973.

3. R. Fossum and B. Iverson, On Picard groups of algebraic fibre spaces, J. Pure and Applied Algebra 3(1973), 269-280.

4. A. Grothendieck, Le groupe de Brauer I,II,III in Dix Exposés sur la Cohomologie des Schémas, North-Holland, Amsterdam, 1968.

5. F. Ischebeck, Zur Picard-Gruppe eines Produktes, Math. Z. 139 (1974), 141-157.

6. A. Magid, The Picard sequence of a fibration, Proc. Amer. Math. Soc. 53(1975).

The Pierce representation of Azumaya algebras

George Szeto

1. Introduction. Let R be a commutative ring with identity 1. G. Azumaya [5] proved that an R-algebra A free as R-module is a central separable algebra if and only if there exists a set of generators $\{a_1, \ldots, a_n\}$ of A such that the matrix $[(a_i a_j)]$ is invertible in A. More characterizations of a central separable (Azumaya)algebra were given by M. Auslander and O. Goldman ([4], Theorem 2.1). Moreover, I. Kaplansky [10] proved that a primitive ring satisfying a polynomial identity with coefficients in the centroid is an Azumaya algebra over the center. This important theorem of Kaplansky was then generalized by M. Artin to an A_n-ring, where a ring A is called an A_n-ring if (1) it satisfies all the identities of n by n matrices, and (2) no homomorphic image of A satisfies the identities of (n-1) by (n-1) matrices ([13], Definition 3.1). Recently, a further generalization was given by C. Procesi [13]. We note that no reference is given to the center of the algebra A in the characterizations of Artin and Procesi. The proof of Procesi was later simplified by S. Amitsur [1]. The purpose of the present paper is to show a characterization in terms of the Pierce sheaf of rings A_x where A_x are stalks of a sheaf induced by a finitely generated R-algebra A (that is, A is finitely generated as a ring over R). Of course, the class of finitely generated R-algebras is larger than that of R-algebras finitely generated as R-modules. It is proved that a finitely generated R-algebra A is an Azumaya R-algebra if and only if so is A_x over R_x for each A_x. Thus a characterization of an Azumaya algebra over a commutative regular ring (in the sense of von Neumann) is derived.

2. Basic definitions. Throughout, we assume that R is a commutative ring with identity 1, that all modules are unitary left modules

over a ring or an algebra and that A is an R-algebra. Let B(R) denote the Boolean algebra of the idempotents of R and SpecB(R) the Boolean space with hull-kernel topology. A system of basic open neighborhoods for this topology are open and closed sets $U_e = \{x$ in SpecB(R) / (1-e) is in $x\}$ for e in B(R). It is known that SpecB(R) is a totally disconnected, compact and Hausdorff topological space. The ring R induces on SpecB(R) a sheaf of rings R_x (= R/xR), called the Pierce sheaf such that R is isomorphic with the ring of global sections of the sheaf ([12] Theorem 4.4). We shall employ some results given in [8], [11], [12] and [14]. Also we shall use the following partition property of SpecB(R) very often: Let $\{U_e\}$ be a cover of SpecB(R). Then there exists a finite cover $\{U_{e_i}$ / i = 1, ..., n$\}$ of SpecB(R) which is a refinement of $\{U_e\}$, where e_i are orthogonal idempotents in B(R) summing to 1. For an R-module M, denote $R_x \otimes_R M$ by M_x, and for an R-algebra A, denote $R_x \otimes_R A$ by A_x.

3. <u>A characterization of Azumaya algebras</u>. In this section, we shall prove the main theorem of this paper: Let R be a commutative ring and A a finitely generated R-algebra. Then A is an Azumaya R-algebra if and only if A_x is Azumaya over R_x for every x in SpecB(R). We begin with three basic lemmas. The first and third are easy to prove and the second is from Corollary of [3].

<u>Lemma</u> 3.1. An Azumaya algebra A over a commutative ring R is finitely presented as an R-algebra.

<u>Lemma</u> 3.2. Azumaya algebras are preserved under central extensions and homomorphic images, where an R-algebra A' is called a central extension of A if there is an R-algebra homomorphism f: A → A', and R' a subalgebra of the center of A', such that A' = f(A)R' ([3]).

<u>Lemma</u> 3.3. If A is an Azumaya algebra over a commutative ring R,

then there exists a finitely generated subring $R_o \subset R$ and an Azumaya R_o-algebra $A_o \subset A$ such that $A = A_oR$.

By the above lemmas, the main theorem (in a local form) can be proved.

Theorem 3.4. Let S be a topological space, R a sheaf of rings on S, A a sheaf of R-algebras, and s a point of S such that (1) A_s is an Azumaya algebra over R_s, and (2) there exists a neighborhood U of s and a finite set of sections $x_1, \ldots, x_m$ in A_U, such that throughout U, A is generated by $\{x_1, \ldots, x_m\}$ as an R-algebra; that is, for each neighborhood $V \subset U$, A_V is generated as an R_V-algebra by $\{(x_i)_V\}$. Then on U', A is a sheaf of Azumaya algebras over R for some neighborhood $U' \subset U$. In particular, $A_{U'}$ is an Azumaya algebra over $R_{U'}$.

Proof. By hypothesis, A_s is an Azumaya R_s-algebra for some s in S, so there exist a finitely generated subring $(R_s)_o$ of R_s and an Azumaya subalgebra $(A_s)_o$ of A_s as in Lemma 3.3. Clearly, we may take $(A_s)_o$ to contain the $(x_i)_s$ for i = 1, ..., m. Note that since $(R_s)_o$ is a finitely generated commutative ring, it will be finitely presented, hence from Lemma 3.1, we see that $(A_s)_o$ is finitely presented as a ring. Let U' be a neighborhood of s contained in U, to which we can lift all elements of a finite generating set for $(R_s)_o$, and of a finite generating set for $(A_s)_o$ over $(R_s)_o$ containing $\{x_i\}$, and in which the set of defining relations for these rings in terms of these generators continues to hold. Let A_o denote the subsheaf of A on U' generated by these elements, a sheaf of homomorphic images of $(A_s)_o$. It is clear from our construction that on U', we have $A = A_oR$, so on U', A is a sheaf of central extensions of homomorphic images of $(A_s)_o$, hence a sheaf of Azumaya algebras over R. The proof is then complete.

Corollary 3.5. Let R be a commutative ring, and A a finitely generated R-algebra. Then A is an Azumaya R-algebra if and only if A_x is an Azumaya R_x-algebra for every x in SpecB(R).

Proof. For the necessity, since each direct summand of A in $(Ae \oplus A(1-e))$ for an e in B(R) is Azumaya and since $A_x = \varinjlim(A/Ae)$ for all e in x, A_x is Azumaya. Conversely, taking U = all of SpecB(R), we apply the above theorem at each point x of SpecB(R) to get a neighborhood U' of x such that $A_{U'}$ is an Azumaya algebra over $R_{U'}$. Since $\{U_e$ for all e in B(R)$\}$ is a system of basic open neighborhoods, U' can be taken to be U_e for some e in B(R). Let x vary over SpecB(R); then SpecB(R) is covered by such U_e's. Thus we refine this cover to get a finite cover, $\{U_{e_i}\}$, where i = 1, ..., n for some integer n, such that e_i are orthogonal idempotents in B(R) summing to 1. Consequently, $A \cong \oplus\sum_{i=1}^{n} Ae_i$ is an Azumaya algebra over $R \cong \oplus\sum_i (Re_i)$, where $Ae_i \cong A_{U_{e_i}}$, and $Re_i \cong R_{U_{e_i}}$ for each i.

In particular, let A be an R-algebra finitely generated as an R-module. We have:

Corollary 3.6. Let R be a commutative ring, and A an R-algebra finitely generated as an R-module. Then A is Azumaya over R if and only if so is A_x over R_x for each x in SpecB(R).

4. Centers of finitely generated algebras. In Theorem 3.4, no reference is given to the center of the finitely generated R-algebra A. We now show that the center of A is determined by the centers of the stalks A_x. Thus a similar characterization of an Azumaya algebra can be obtained. By using the sheaf technique as given in section 3, we have:

Lemma 4.1. Let R be a commutative ring, A a finitely generated

R-algebra, and R' the subalgebra of A. Then R' is the center of A if and only if $(R')_x$ is the center of A_x for each x in SpecB(R).

Theorem 4.2. Let R be a commutative ring, and A a finitely generated R-algebra such that for each x in SpecB(R), A_x is Azumaya over its center. Then A is Azumaya over its center.

Corollary 4.3. Let A be a biregular ring with center R. Then A is a finitely generated R-module if and only if it is an Azumaya R-algebra.

5. Examples. The following examples are given to illustrate (1) the Azumaya algebra A of Theorem 3.4, and (2) the subalgebras of A which are not Azumaya, but with different centers.

Let S be the set $\{1, 1/2, \ldots, 0\}$, and let R be the ring of all locally constant functions from S into a field K.

Let A_1 denote the 2 by 2 matrix ring over R, $M_2(R)$; equivalently, the algebra of locally constant functions from S into $M_2(K)$. This is, of course, Azumaya; for each x in S, $(A_1)_x = M_2(K)$.

Now define $A_2 \subset A_3 \subset A_1$ to be the subalgebras consisting of all elements a such that a_0 is scalar (of the form $\begin{bmatrix} \alpha, 0 \\ 0, \alpha \end{bmatrix}$) in the case of A_2, and diagonal (of the form $\begin{bmatrix} \alpha, 0 \\ 0, \beta \end{bmatrix}$) in the case of A_3. Neither of these are finitely generated, because for any finitely generated subalgebra A, there will exist a neighborhood of the origin in which all elements have scalar or diagonal stalks. The stalks of A_2 and A_3 are still $M_2(K)$ at all points except 0.

Clearly, A_2 has the property that all stalks are Azumaya over R_x, yet it is not Azumaya over R, since it is not finitely generated. A_3 has the property that its stalk at the point 0 has center larger than R_x, though R = center(A_3).

References

1. S. Amitsur, Polynomial identities and Azumaya algebras, J. Alg. 27(1973), 117-125.

2. R. Arens and I. Kaplansky, Topological representation of algebras, Trans. Amer. Math. Soc. 63(1948), 457-481.

3. M. Artin, On Azumaya algebras and finite dimension representation of rings, J. Alg. 11(1969), 532-563.

4. M. Auslander and O. Goldman, The Brauer group of a commutative ring, Trans. Amer. Math. Soc. 97(1960), 367-409.

5. G. Azumaya, On maximality central algebras, Nagoya J. Math. 2 (1951), 119-150.

6. H. Bass, Lectures on topics in algebraic K-theory, Tata Institute of Fundamental Research, Bombay, 1967.

7. G. Bergman, Hereditary commutative ring and centres of hereditary rings, Proc. London Math. Soc. 23(1971), 214-236.

8. J. Dauns and K. Hofmann, The representation of biregular rings by sheaves, Math. Zeitschr. 91(1966), 103-123.

9. F. DeMeyer and E. Ingraham, Separable algebras over commutative rings, Springer-Verlag, Berlin-Heidelberg-New York, 1971.

10. I. Kaplansky, Rings with a polynomial identity, Bull. Amer. Math. Soc. 54(1948), 575-580.

11. A. Magid, Pierce's representation and separable algebras, Ill. J. Math. 15(1971), 114-121.

12. R. Pierce, Modules over commutative regular rings, Mem. Amer. Math. Soc. 70, 1967.

13. C. Procesi, On a theorem of M. Artin, J. Alg. 22(1972), 309-315.

14. O. Villamayor and D. Zelinsky, Galois theory for rings with infinitely many idempotents, Nagoya J. Math. 35(1969), 359-368.

A REMARK ON CLASS GROUPS

William H. Gustafson

§1. Introduction

Let R be a Dedekind domain with field of fractions K, and let Λ be an R-order in a separable K-algebra A. We sketch a description of the locally free class group of Λ; for details see [5]. Two left Λ-modules M and M' are called stably isomorphic if there is a finitely generated free left Λ-module F such that $M \oplus F \cong M' \oplus F$. A fractional left Λ-ideal M in A is called locally free if $M_P \cong \Lambda_P$ for all prime ideals P of R. (Throughout, the subscript P denotes completion at P.) If M and M' are locally free Λ-ideals, then $M \oplus M'$ admits a decomposition $M \oplus M' \cong \Lambda \oplus M''$, where M'' is a locally free Λ-ideal whose stable isomorphism class is determined by those of M and M'. Using this fact, one can construct an abelian group $C(\Lambda)$, the locally free class group of Λ, whose elements are the stable isomorphism classes $[M]$ of locally free Λ-ideals. The addition is given by $[M]+[M'] = [M'']$ if $M \oplus M' \cong \Lambda \oplus M''$. If Λ is commutative, stable isomorphism implies isomorphism, and in the addition formula above, we may take

$$M'' = M \cdot M' = \{\Sigma x_i y_i \mid x_i \in M,\ y_i \in M'\},$$

where the sums are computed inside A. Hence, $C(\Lambda)$ may be identified with the Picard group $\mathrm{Pic}_\Lambda(\Lambda)$. In particular, $C(R)$ is the usual ideal class group of R.

If Λ' is a maximal R-order in A containing Λ, there is a surjection $\theta : C(\Lambda) \to C(\Lambda')$, given by $\theta([M]) = [\Lambda' \otimes_{\Lambda} M]$. Denoting $\ker \theta$ by $D(\Lambda)$, we have an exact sequence

$$(*) \qquad 0 \to D(\Lambda) \to C(\Lambda) \to C(\Lambda') \to 0 .$$

The group $D(\Lambda)$ is determined wholly by Λ; it does not depend on the choice of the maximal order Λ' containing Λ. In most cases, the computation of $C(\Lambda')$ reduces to the determination of various ray class groups of Dedekind domains. Hence, attention has focused on the kernel group $D(\Lambda)$. The calculation of $C(\Lambda)$ also requires knowledge of the structure the extension (*). Ullom [6] has shown that (*) does not split when Λ is the integral group ring of a cyclic group of order p^n, where p is a properly irregular prime and $n \geq 2$. In this note, we construct some simple examples of cases where (*) splits and does not split, using orders in imaginary quadratic fields. Our calculations are based on a class number formula due to Dedekind, which we derive here from general results of Fröhlich.

§2. Orders in quadratic fields

Let d be a square-free rational integer, we will consider the field $A = Q(d^{\frac{1}{2}})$. There is a unique maximal $\mathbb{Z}$-order in A, namely $\Lambda' = \mathbb{Z}[\omega]$, where

$$\omega = \begin{cases} d^{\frac{1}{2}}, & \text{if } d \not\equiv 1 \pmod 4 \\ (1 + d^{\frac{1}{2}})/2, & \text{if } d \equiv 1 \pmod 4 . \end{cases}$$

Any $\mathbb{Z}$-order Λ in A has the form $\mathbb{Z}[n\omega]$, where $n = [\Lambda' : \Lambda]$ is the <u>conductor</u> [1, p.48]. For any ring R, let $Y(R)$ denote the group of units of R. Let h_A denote the class number of A, and for primes p of $\mathbb{Z}$, define the Kronecker symbol $\left(\frac{A}{p}\right)$ by

$$\left(\frac{A}{p}\right) = \begin{cases} 1\ , & \text{if } p \text{ splits in } A \\ -1\ , & \text{if } p \text{ remains prime in } A \\ 0\ , & \text{if } p \text{ ramifies in } A \end{cases}$$

We want to prove Dedekind's formula [2] :

$$|C(\Lambda)| = [Y(\Lambda') : Y(\Lambda)]^{-1} \cdot n \cdot h_A \cdot \prod_{p|n} (1 - (\tfrac{A}{p})p^{-1})\ .$$

The formula is usually proved by a study of zeta functions [1]; we will obtain it from an exact sequence of Fröhlich [3]. His result is valid for an arbitrary commutative order Λ over a Dedekind ring R. Letting Λ' denote the unique maximal R-order in $K \otimes_R \Lambda$, he obtains the exactness of

$$0 \to Y(\Lambda')/Y(\Lambda) \to \prod_{S(\Lambda)} Y(\Lambda'_p)/Y(\Lambda_p) \to C(\Lambda) \to C(\Lambda') \to 0\ ,$$

where $S(\Lambda) = \{P \in \operatorname{spec} R \mid \Lambda_p \neq \Lambda'_p\}$. The proof is by idelic methods.

Clearly Dedekind's formula follows from the exactness of this sequence if we show

$$[Y(\Lambda'_p) : Y(\Lambda_p)] = (1 - (\tfrac{A}{p})p^{-1}) \cdot p^{\nu_p(n)}\ ,$$

where $p^{\nu_p(n)} \| n$, $p^{\nu_p(n)} \geq p$. Write $[Y(\Lambda'_p) : Y(\Lambda_p)] = \ell_p \cdot m_p$, where m_p is a power of p and $(\ell_p, p) = 1$. Let $\mathfrak{R}_p$ denote the Jacobson radical of Λ'_p. For ideals $\mathfrak{p}$, $\mathfrak{q}$ of Λ' and Λ

respectively, let $N(\mathfrak{p})$, $N(\mathfrak{q})$ denote the norms $[\Lambda' : \mathfrak{p}]$ and $[\Lambda : \mathfrak{q}]$. Fröhlich shows

$$\ell_p = \frac{|Y(\Lambda_p'/R_p)|}{|Y(\Lambda_p/R_p \cap \Lambda_p)|} = \frac{\prod (N\mathfrak{p}_j - 1)}{\prod (N\mathfrak{q}_k - 1)}$$

where $\mathfrak{p}_j$ runs over the maximal ideals of Λ_p' , and $\mathfrak{q}_k$ over those of Λ_p . Further,

$$m_p = [R_p : R_p \cap \Lambda_p] = [\Lambda_p' : \Lambda_p] \cdot [\Lambda_p : R_p \cap \Lambda_p]/[\Lambda_p' : R_p] .$$

Hence, it suffices to show

$$m_p = \begin{cases} p^{\nu_p(n)-1}, & \text{if } p \text{ does not ramify in } A \\ p^{\nu_p(n)}, & \text{if } p \text{ ramifies in } A \end{cases}$$

and

$$\ell_p = \begin{cases} p-1, & \text{if } p \text{ splits in } A \\ p+1, & \text{if } p \text{ remains prime in } A \\ 1, & \text{if } p \text{ ramifies in } A \end{cases}$$

Most of this will be deduced from

Proposition: *Let* $p|n$. *Then* Λ_p *is a local ring with maximal ideal* $\mathfrak{m} = \mathbb{Z}_p \cdot p \oplus \mathbb{Z}_p \cdot n\omega$ *and residue field isomorphic to* $\mathbb{Z}/(p)$.

Proof: Consider for instance the case where $\omega^2 = d < 0$. Then $(n\omega)^2 = n^2 d$, a multiple of p , whence it follows easily that $\mathfrak{m}$ is an ideal of Λ_p . Let $x = (\alpha/\beta) + (\mu/\nu)n\omega \in \Lambda \setminus \mathfrak{m}$, where the fractions are in lowest terms. Then $p \nmid \alpha\beta\nu$ and $x\bar{x} = ((\alpha\nu)^2 -$

$(\beta\mu n)^2 d)/(\beta\nu)^2 \in Y(\mathbb{Z}_p)$, whence x is a unit of Λ_p . The other cases are similar.

The values of ℓ_p announced above can now be verified. For example, if p splits in A , then Λ'_p has two maximal ideals P_1 and P_2 , each of norm p , while by the proposition, Λ_p has a unique maximal ideal $\mathfrak{m}$ of norm p . Hence Fröhlich's formula yields

$$\ell_p = (NP_1 - 1)(NP_2 - 1)(N\mathfrak{m} - 1)^{-1}$$

$$= (p-1)(p-1)(p-1)^{-1} = p-1 .$$

To calculate m_p , we first note that $[\Lambda'_p : \Lambda_p] = p^{\nu_p(n)}$. Also, $\Lambda_p \cap \mathfrak{R}_p$ is the unique maximal ideal of Λ_p , so $[\Lambda_p : \Lambda_p \cap \mathfrak{R}_p] = p$. Since it is clear that $[\Lambda'_p : \mathfrak{R}_p]$ equals p if p ramifies in A , and equals p^2 otherwise, Fröhlich's formula yields the values announced above.

§3. Examples

In this section, we will examine some orders in $A = Q(\omega)$, where $\omega = (-5)^{\frac{1}{2}}$. Let Λ' denote the unique maximal order $\mathbb{Z}$-order $\mathbb{Z}[\omega]$ in A . Note that $C(\Lambda')$ has order 2; the nontrivial ideal class is represented by $\Lambda' \cdot (1 + 2\omega) + \Lambda' \cdot 3$.

For our first example, we take $\Lambda = \mathbb{Z}[2\omega]$. Let $P = \mathbb{Z} \cdot (1 + 2\omega) \oplus \mathbb{Z} \cdot 6\omega$, an ideal of Λ . If p is a rational prime, we have $P_p = \Lambda_p$, save for the case $p = 3$, when $P_p = \Lambda_p \cdot (1 + 2\omega)$. Thus, P is locally free. We will show that $[P]$ has order four in $C(\Lambda)$. Since $|C(\Lambda)| = 4$ by Dedekind's formula, it follows then that (*) does not split. An easy calculation shows that $P^2 = \mathbb{Z} \cdot 18\omega \oplus \mathbb{Z} \cdot (1 - 4\omega)$, whence Λ/P^2 has nine elements. If $x = \alpha + \beta \cdot (2\omega) \in \Lambda$, then $\Lambda/\Lambda x$ has $\alpha^2 + 20\beta^2$ elements. Hence, if P^2 is principal, then $P^2 = \Lambda \cdot 3$. But $3 \notin P^2$, whence P^2 is nonprincipal, and $[P]^2$ is not the trivial element of $C(\Lambda)$. We remark that $\Lambda'P = \Lambda' \cdot (1+2\omega) + \Lambda' \cdot 3$, $\Lambda'P^2 = \Lambda' \cdot (2+\omega)$, and $P^4 = \Lambda \cdot (1-4\omega)$.

In our second example, we take $\Lambda = \mathbb{Z}[3\omega]$, and Dedekind's formula shows that once again, we have $|C(\Lambda)| = 4$. We will show that (*) does split in this case, so that $C(\Lambda)$ is a Klein group. For this, it suffices to find an element of order two of $C(\Lambda)$ whose image in $C(\Lambda')$ also has order two. Let P be the ideal of Λ having $\mathbb{Z}$-basis $\{1+3\omega, 2\}$. We find that $P_p = \Lambda_p$ for p odd, while $P_2 = \Lambda_2 \cdot (1+3\omega)$. Further, $P^2 = \Lambda \cdot 2$, and $\Lambda'P$ is the nonprincipal ideal of Λ' generated by 2 and $1+\omega$. This completes the verification.

Finally, it is easy to produce many cases where (*) splits for trivial reasons. For instance, for $\Lambda = \mathbb{Z}[5\omega]$, we find $|D(\Lambda)| = 5$, whence (*) splits.

4. Concluding remarks

One would like to know necessary and sufficient conditions for the splitting of (*), as a first step toward determining the structure of $C(\Lambda)$. There is such a condition, but it seems impossible to verify in most cases. To formulate it, we use a formula of Jacobinski [4]. Let R be a Dedekind domain with field of fractions K , A a separable K-algebra and Λ an R-order in A . Let Λ' be a maximal order containing Λ , and let $\mathfrak{f}$ be a nonzero ideal of R such that $\Lambda' \cdot \mathfrak{f} \subseteq \Lambda$. Let F denote the center of A , C the integral closure of R in F , and $N: A \to F$ the reduced norm. We define $R_{\mathfrak{f}} = \bigcap_{P|\mathfrak{f}} R_P$, $\Lambda_{\mathfrak{f}} = R_{\mathfrak{f}} \otimes_R \Lambda$. An element of Λ is said to be <u>prime to $\mathfrak{f}$</u> if it is a unit in $\Lambda_{\mathfrak{f}}$.

Let $I(C, \mathfrak{f})$ be the multiplicative group of full, fractional C-ideals in F which are prime to $\mathfrak{f}$. Define subgroups $I(\Lambda)$ and $I(\Lambda')$ of $I(C, \mathfrak{f})$ as those generated by $\{C \cdot N(x) \mid x \in \Lambda,$ x prime to $\mathfrak{f}\}$ and $\{C \cdot N(x) \mid x \in \Lambda',$ x prime to $\mathfrak{f}\}$, respectively. Then we have a commutative diagram

$$\begin{array}{ccc} C(\Lambda) & \xrightarrow{\ \theta\ } & C(\Lambda') \\ \downarrow \cong & & \downarrow \cong \\ I(C,\mathfrak{f})/I(\Lambda) & \longrightarrow & I(C,\mathfrak{f})/I(\Lambda') \end{array} .$$

We claim that the surjection θ splits if and only if there is an endomorphism s of $I(C, \mathfrak{f})$ such that $s(I(\Lambda')) \subseteq I(\Lambda)$ and $\operatorname{Im}(s - \text{identity}) \subseteq I(\Lambda')$. Thus, we need a sort of norm from Λ' to Λ. I know of no general procedure for finding such maps. The criterion for splitting follows from the proposition below, whose easy proof is omitted.

Proposition: Let $A \subseteq B \subseteq P$ be left modules over any ring, with P projective. Then the natural surjection $P/A \to P/B$ splits if and only if P has an endomorphism s such that $s(B) \subseteq A$ and $\operatorname{Im}(s - 1_P) \subseteq B$.

REFERENCES

1. H. Cohn: A second course in number theory, John Wiley & Sons, Inc., New York, 1962.

2. R. Dedekind: Über die Anzahl der Ideal-Klassen in den verschiedenen Ordnungen eines endlichen Körpers, Festschrift der Technischen Hochschule in Braunschweig zur Säkularfeier des Geburtstages von C. F. Gauss, Braunschweig, 1877, 1-55; also Gesammelte mathematische Werke I, 105-157.

3. A. Fröhlich: On the classgroup of integral grouprings of finite abelian groups, Mathematika 16(1969), 143-152.

4. H. Jacobinski: Genera and decompositions of lattices over orders, Acta Math. 121(1968), 1-29.

5. I. Reiner and S. Ullom: Class groups of integral group rings, Trans. Amer. Math. Soc. 170(1972), 1-30.

6. S. Ullom: The Δ-decomposition of the class group of cyclic p-groups, Notices Amer. Math. Soc. 23(1976), A-57 - A-58.

SPLITTING OF AZUMAYA ALGEBRAS OVER NUMBER RINGS

Philip La Follette

Let K be an algebraic number field, of finite degree over the rational field Q ; let L be a cyclic extension field of K , with $n = [L:K]$. We denote by R and S the rings of integers of K and L respectively. Let E be a finite set of prime spots of K that includes all the infinite prime spots, and let E' be the set of all prime spots of L that lie above members of E . We denote by K^E the (multiplicative) group of "E-units" of K , i.e., the set of all members of K that are units at all the primes outside E . Likewise we denote by $L^{E'}$ the group of E'-units of L .

The norm $N_{L/K}$ from L to K provides a homomorphism from $L^{E'}$ into K^E . In [3, §7] Chevalley calculated that, if E is large enough, and if for each prime spot $\mathfrak{p}$ in E we denote by $n_{\mathfrak{p}}$ the degree of $L\hat{K}_{\mathfrak{p}}$ over the completion $\hat{K}_{\mathfrak{p}}$ of K at $\mathfrak{p}$, then

$$(1) \qquad (K^E : N_{L/K}(L^{E'})) = (\prod n_{\mathfrak{p}})/n ,$$

the product taken over all $\mathfrak{p}$ in E . The calculation used the Herbrand quotient and Minkowski's lemma on the existence of lattice points in parallelotopes.

In this note we interpret the quotient group $K^E/N_{L/K}(L^{E'})$ as a Brauer group. We denote by R^E the Dedekind domain consisting of those members of K having no poles except possibly at members of E ; thus R^E is the intersection of those valuation rings in K that correspond to primes outside E . Similarly we denote by $S^{E'}$ the ring of members of L having no poles outside E' . Then $S^{E'}$ is the integral closure of R^E in L . The Galois group G of L over K acts on $S^{E'}$, and it is not hard to see that $S^{E'}$ is a Galois extension of R^E (in the sense of Chase, Harrison, and Rosenberg [2]) if and only if E contains all the prime spots of K that are ramified in the extension L/K . The groups K^E and $L^{E'}$ are the groups of units of R^E and $S^{E'}$ respectively.

THEOREM. If E contains all the prime spots of K that are ramified in the extension L/K , and if E is large enough that the order of $\mathrm{Pic}(S^{E'})$ is relatively prime to n , then

$$K^E/N_{L/K}(L^{E'}) \cong \mathrm{Br}(S^{E'}/R^E) .$$

Proof. Under the hypothesis that E contains the ramified prime spots, the ring $S^{E'}$ is a Galois extension of R^E . Thus there is a Chase-Rosenberg exact sequence

$$H^0(G,\mathrm{Pic}(S^{E'})) \xrightarrow{\alpha} H^2(G,L^{E'}) \xrightarrow{\beta} \mathrm{Br}(S^{E'}/R^E) \xrightarrow{\gamma} H^1(G,\mathrm{Pic}(S^{E'}))$$

of finite abelian groups. Since G is cyclic, the group $H^2(G,L^{E'})$ is naturally isomorphic to $K^E/N_{L/K}(L^{E'})$. Moreover this group is annihilated by n , since every n^{th} power in K^E is a norm. The Brauer group $\mathrm{Br}(S^{E'}/R^E)$ is embedded in the Brauer group Br(L/K) (since R^E is a Dedekind domain and K its quotient field), and so is also annihilated by n . It follows, since by hypothesis the order of $\mathrm{Pic}(S^{E'})$ is prime to n , that the homomorphisms α and γ are zero, and β is an isomorphism, Q. E. D.

The group $\mathrm{Pic}(S^{E'})$ is isomorphic to the quotient of $\mathrm{Pic}(S)$ by the subgroup generated by the finite primes belonging to E' [4, Theorem 7.1].

COROLLARY. Let D be the discriminant of L (over Q). If E contains all the prime spots of K that are ramified in the extension L/K, and also all those (finitely many) finite prime spots $\mathfrak{p}$ of K such that $N_{K/Q}(\mathfrak{p}) \leq |D|^{1/2n}$, then $K^E/N_{L/K}(L^{E'})$ is isomorphic to $\mathrm{Br}(S^{E'}/R^E)$.

Proof. Under the hypothesis, $\mathrm{Pic}(S^{E'})$ is trivial.

EXAMPLE. Let Z be the ring of rational integers. We can use the Theorem to show, without using the Takagi-Artin reciprocity law, that certain subgroups of $\mathrm{Br}(Z)$ are trivial. Let $K = Q$, and $L = Q(\sqrt{m})$. Suppose that m is a positive squarefree integer such that the discriminant of L is prime; in other words, m is a prime congruent to 1 modulo 4. Then m is the only rational prime ramified in L, and if E is the set whose two elements are the infinite prime spot of Q and the prime spot of Q corresponding to m, the formula (1) shows that $Q^E/N_{L/Q}(L^{E'})$ is trivial. Moreover $\mathrm{Pic}(S)$ has odd order [1, p. 247, Corollary], and so its homomorphic image $\mathrm{Pic}(S^{E'})$ also has odd order. Therefore, by the Theorem, $\mathrm{Br}(S^{E'}/Z^E)$ is trivial. Since $\mathrm{Br}(S/Z)$ is embedded naturally in $\mathrm{Br}(S^{E'}/Z^E)$, we conclude that $\mathrm{Br}(S/Z)$ is trivial.

REFERENCES

[1] Z. I. Borevich and I. R. Shafarevich, Number Theory, Academic Press, New York, 1966 (translation).

[2] S. U. Chase, D. K. Harrison, and A. Rosenberg, Galois Theory and Galois Cohomology of Commutative Rings, Mem. Amer. Math. Soc., No. 52 (1965), 15-32.

[3] C. Chevalley, La théorie du corps de classes, Ann. of Math. 41 (1940), 394-418.

[4] R. M. Fossum, The Divisor Class Group of a Krull Domain, Ergebnisse der Mathematik und ihrer Grenzgebiete, Springer-Verlag, New York, 1973.

Abelian p-extensions and cohomology

H. F. Kreimer

1. Preliminaries: Throughout this report ring (algebra) will mean ring (algebra) with identity element, usually denoted by 1; and for any set X, 1_X will denote the identity map on X. Let R be a given commutative ring; and let J be a Hopf algebra over R with multiplication map $\pi: J \otimes J \to J$, unit map $i : R \to J$, comultiplication map $\psi: J \to J \otimes J$, counit map $\varepsilon : J \to R$, and antipode $\lambda : J \to J$. Assume that J is a finitely generated, projective R-module; and let J^* denote the dual Hopf algebra, the elements of which are the R-module homomorphisms of J into R. The following definition is a straightforward extension of Definition 7.3 of [3] to not necessarily commutative aglebras.

Definition: An R-algebra S will be called J-Galois if: (1) S is a faithfully flat R-module; (2) there exists an R-algebra homomorphism $\alpha : S \to J \otimes S$, mapping identity element onto identity element, for which $(1_J \otimes \alpha) \circ \alpha = (\psi \otimes 1_S) \circ \alpha$ and $(\varepsilon \otimes 1_S) \circ \alpha = 1_S$; (3) the unique extension of α to a right S-module homomorphism $\gamma : S \otimes S \to J \otimes S$ is an isomorphism.

Remarks: The algebra J is J-Galois with respect to the comultiplication map $\psi : J \to J \otimes J$. In this case, the map γ is $(1_J \otimes \pi) \circ (\psi \otimes 1_J)$ and its inverse is $(1_J \otimes \pi) \circ (1_J \otimes \lambda \otimes 1_J) \circ (\psi \otimes 1_J)$. Also, if S is any R-algebra which is J-Galois, then just as in Theorem 9.3 of [3] it is easily shown that S is a finitely generated, projective R-module and the left S-module $\mathrm{Hom}_R(S, S)$ of R-module endomorphisms of S is a smash or crossed product of S of J^*.

Let G be a finite group and let $R[G]$ denote the group algebra of G over R. Then $R[G]$ is a Hopf algebra with antipode and G is a set of free generators or basis for the R-module $R[G]$. Letting $\{V_\sigma | \sigma \in G\}$ be the dual basis for $R[G]^*$, $\{V_\sigma | \sigma \in G\}$ is a set of pairwise orthogonal idempotent elements whose seem is the identity element of $R[G]^*$.

Example 1: Let $J = R[G]^*$. Then an R-algebra S is J-Galois if and only if the following conditions are satisfied:

(1) There is a representation of G as a group of automorphisms of the R-algebra S.

(2) S is a faithful R-algebra and R is the subring of G-invariant elements of S.

(3) There exist elements $x_1, \ldots, x_n; y_1, \ldots, y_n$ of S such that $\sum_{i=1}^{n} \sigma(x_i) \cdot y_i = \delta_{\sigma,1}$ for σ in G.

Example 2: Let $J = R[G]$. Then an R-algebra S is J-Galois if and only if the following conditions are satisfied:

(1) For each σ in G there exists a faithfully flat R-submodule K_σ of S, such that the R-module S is the direct sum $\sum_{\sigma \in G} \oplus K_\sigma$.

(2) $K_\sigma \cdot K_\tau \subseteq K_{\sigma\tau}$ for σ, τ in G.

(3) The restriction of the multiplication map of S is an isomorphism of $S \otimes K_\sigma$ onto S for each σ in G.

Moreover, if the R-algebra S is J-Galois, then V_σ acts as the projection of S onto K_σ for each σ in G.

Remark: Let R be a field of positive characteristic p. If S is a simple, purely inseparable field extension of dimension p^e over R, Say $S = R[x]$, and $G = (\sigma)$ is a cyclic group of order p^e; then S is seen to be R[G]-Galois by setting $K_{\sigma i} = R \cdot x^i$ for $0 \leq i < p^e$. More generally, because a finite dimensional, modular, purely inseparable field extension of R is isomorphic to a tensor product of simple field extensions of R, it is R[G]-Galois for a group G which is a direct product of cyclic p-groups.

Finally let G be a finite abelian group, and let S be a commutative R-algebra which is R[G]-Galois. If G_1 is a subgroup of G and $T = \sum_{\sigma \in G_1} K_\sigma$, then it is readily verified that T is a subalgebra of S which is $R[G_1]$-Galois and S is $T[G/G_1]$-Galois as an algebra over T.

2. Amitsur and Harrison cohomologies. Let S be a commutative R-algebra, let G be a finite ableian group, and let $J = R[G]$. The Amitsur cohomology arises from the cosimplicial algebra $C(S/R): S \rightrightarrows S^2 \substack{\rightarrow\\\rightarrow\\\rightarrow} S^3 \ldots$, where S^k denotes the k-fold tensor product of S with itself and the face operators $d_i: S^k \to S^{k+1}$ are specified by the equations $d_i(x_1 \otimes \ldots \otimes x_k) = x_1 \otimes \ldots \otimes x_{k-i} \otimes 1 \otimes x_{k-i+1} \ldots \otimes x_k$, the x_j being arbitrary elements of S and $0 \leq i \leq k$. If S is J-Galois, then by iteration of the isomorphism $\gamma: S \otimes S \to J \otimes S$ an identification of S^k with $J^{k-1} \otimes S$ is obtained for each positive integer k (with the understanding that $J^0 = R$). The face operators from $S^k = J^{k-1} \otimes S$ into $S^{k+1} = J^k \otimes S$ may then be described as follows: $d_k(z) = 1 \otimes z$ for z in $J^{k-1} \otimes S$; d_i is the comultiplication map ψ applied to the $(k-i)$-th factor J for $0 < i < k$; and d_0 is the map $\alpha: S \to J \otimes S$ applied to the last factor S. By restricting these face operators, a cosimplicial subalgebra $A(R,G): R \rightrightarrows J \substack{\rightarrow\\\rightarrow\\\rightarrow} J^2 \ldots$ is obtained.Note that at the first term d_0 and d_1 restrict to the unit map $i: R \to J$, and d_0 restricts to the map $z \leadsto z \otimes 1$ for z in J^k and $k \geq 1$.

A functor L from the category of commutative R-algebras to the category of abelian groups will carry the cosimplicial algebras $C(S/R)$ and $A(R,G)$ to cosimplicial abelian groups, and these become cochain complexes $C(S/R,L)$ and $A(R,G,L)$, respectively, with alternating sums of face operators as boundary homomorphisms. Two such functors are the forgetful functor F which assigns to an R-algebra its additive group and the units functor U which assigns to an R-algebra its multiplicative group of invertible elements. For $n \geq 0$, the Amitsur cohomology groups $H^n(S/R,U)$ are the cohomology groups of the complex $C(X/R,U)$ and the Harrison cohomology groups $H^n(R,G,U)$ are the cohomology groups of the complex $A(R,G,U)$. Moreover, if S is $R[G]$-Galois, then $A(R,G,U)$ is a subcomplex of $C(S/R,U)$ and therefore there is a canonical homomorphism of $H^n(R,G,U)$ into $H^n(S/R,U)$ for $n \geq 0$. According to [2, Corollary 4.8], the elements of $H^2(R,G,U)$ are in one-to-one correspondence with the isomorphism classes of those Galois extensions of R with Galois group G which have normal bases; and from [2, Definition 4.5 and Remark following Definition 4.5] it is easily seen that a Galois extension of R with Galois group G is just an R-algebra which is $R[G]^*$-Galois. Of course $H^2(S/R,U)$ is closely connected to the relative Brauer group $B(S/R)$ which consists of equivalence classes of central separable R-algebras which are split by S [1, Theorem 7.6].

Finally, if T is a commutative R-algebra, then $T \otimes S$ is a T-algebra and the k-fold tensor product of $T \otimes S$ with itself over T may be identified with $T \otimes S^k$ for $k \geq 1$. Moreover, if $d_i: S^k \to S^{k+1}$ is the i-th face operator of the cosimplicial algebra $C(S/R)$, then $1_T \otimes d_i: T \otimes S^k \to T \otimes S^{k+1}$ is the i-th face operator of the cosimplicial algebra $C(T \otimes S/T)$. If T is a subalgebra of S, then the restriction of the multiplication map of S is a T-algebra homomorphism of $T \otimes S$ onto S and it determines a homomorphism from the cosimplicial algebra $C(T \otimes S/T)$ into the cosimplicial algebra $C(S/T)$. Now suppose that S is $R[G]$-Galois, let G_1 be a subgroup of G, let $J_1 = R[G_1]$, and let T be the corresponding subalgebra of S which is J_1-Galois. For $k \geq 1$, $T \otimes S^k \approx (T \otimes T) \otimes_T S^k \approx (J_1 \otimes T) \otimes_T S^k \approx J_1 \otimes S^k \approx J_1 \otimes J^{k-1} \otimes S$; and by restricting the face operators of $C(T \otimes S/T)$, a cosimplicial algebra $B(R,G:G_1): J_1 \rightrightarrows J_1 \otimes J \rightrightarrows J_1 \otimes J^2 \ldots$ is obtained. Furthermore the restriction of the homomorphism from $C(T \otimes S/T)$ into $C(S/T)$ is a homomorphism from the cosimplicial algebra $B(R,G:G_1)$ into the cosimplicial algebra $A(R,G/G_1)$. In keeping with previous notation, let $B(R,G:G_1,L)$ denote the cochain complex associated with $B(R,G:G_1)$ by a functor L from the category of commutative R-algebras to the category of abelian groups, and let the cohomology groups of $B(R,G:G_1,L)$ be denoted by $H^n(R,G:G_1,L)$ for $n \geq 0$.

3. Theorems of Berkson and Hoechsmann: From hereon assume that R has prime characteristic p, and for the present assume that G is a cyclic group of order p and S is a commutative R-algebra which is R[G]-Galois. Then $D = \sum_{i=1}^{p-1} i\, v_{\sigma^i}$ acts as a derivation of the R-algebra S and $D^p = D$. Since $R[G]$ is R[G]- Galois, D also acts as a derivation of $R[G]$. Use the derivation D to trace the argument of Zelinsky's proof of Berkson's Theorem [7]. Extend D to a derivation of the k-fold tensor product S^k, $k \geq 1$, by letting D act on the first factor of the tensor product. By omitting the first term of the cochain complex $C(S/R,U)$, a cochain complex $C_1(S/R,U)$ is obtained. From the subcomplex $A(R,G,U)$ of $C(S/R,U)$ is obtained likewise a subcomplex $A_1(R,G,U)$ of $C_1(S/R,U)$. Considering $S \otimes S$ to be an S-algebra via the first factor of the tensor product, chain transformations $\lambda: C_1(S/R,U) \to C(S \otimes S/S,F)$ and $\nu: C(S \otimes S/S,F) \to C(S/R,F)$ are defined by the rules $\lambda(x) = (-1)^q \cdot x^{-1} \cdot D(x)$ for x in $U(S^q)$ and $\nu(y) = D^{p-1}(y)+y^p-y$ for y is S^q, $q \geq 2$. Chain transformations $\lambda': A_1(R,G,U) \to B(R,G:G,F)$ and $\nu': B(R,G:G,F) \to A(R,G,F)$ are obtained by restricting λ and ν, respectively. The map ν differs from the corresponding map defined in [7] because the derivation D above is not nilpotent but satisfies $D^p - D = 0$. Still the argument in [7] is easily adapted to show that the rows of the following commutative diagram are exact.

$$\begin{array}{ccccccccc} 0 \to \operatorname{Ker}\lambda & \to & C_1(S/R,U) & \overset{\lambda}{\to} & C(S\otimes S/S,F) & \overset{\nu}{\to} & C(S/R,F) & \to & 0 \\ \uparrow & & \uparrow & & \uparrow & & \uparrow & & \\ 0 \to \operatorname{Ker}\lambda' & \to & A_1(R,G,U) & \overset{\lambda'}{\to} & B(R,G:G,F) & \overset{\nu'}{\to} & A(R,G,F) & \to & 0 \end{array}$$

The cochain complexes of the diagram, except for $C_1(S/R,U)$ and $A_1(R,G,U)$, are found to be acyclic. Therefore, by decomposing the rows into short exact sequences, it can be shown that $H^{n+1}(S/R,U) \approx H^n(\text{Image } \lambda)$ and $H^{n+1}(R,G,U) \approx H^n(\text{Image } \lambda')$ for $n \geq 1$, while $H^n(\text{image } \lambda) = 0 = H^n(\text{Image } \lambda')$ for $n \geq 2$. Also the following commutative diagram of cohomology groups has exact rows.

$$\begin{array}{ccccccc} H^0(S\otimes S/S,F) & \overset{\nu}{\to} & H^0(S/R,F) & \to & H^1(\text{Image } \lambda) & \to & 0 \\ \uparrow & & \uparrow & & \uparrow & & \\ H^0(R,G:G,F) & \overset{\nu'}{\to} & H^0(R,G,F) & \to & H^1(\text{Image } \lambda') & \to & 0 \end{array}$$

But $H^0(S\otimes S/S,F) = S$ and $H^0(S/R,F) = H^0(R,G,F) = H^0(R,G:G,F) = R$. Therefore, $H^1(\text{Image } \lambda) = F(R)/\{D^{p-1}(x)+x^p-x \mid x\in S\}$, $H^1(\text{Image } \lambda') = F(R)/\{x^p-x \mid x\in R\}$ and $H^1(\text{Image } \lambda)$ is a homomorphic image of $H^1(\text{Image } \lambda')$.

<u>Theorem 1</u>. Let R have prime characteristic p, let G be a group of order p, and let S be a commutative R-algebra which is R[G]-Galois. Then $H^n(R,G,U) = 0 = H^n(S/R,U)$ for $n \geq 3$, and $H^2(S/R,U)$ is a homomorphic image of $H^2(R,G,U)$.

This theorem remains true of G is replaced by any finite, abelian p-group,; and a proof by induction on the order of G is possible through the use of the following result.

<u>Theorem 2</u>. Let R have prime characteristic p, let G be a finite, abelian p-group, and let S be a commutative R-algebra which is R[G]-Galois. If G_1 is a subgroup of G and T is the corresponding subalgebra of S which is $R[G_1]$-Galois, then there is a commutative diagram with exact rows:

$$\begin{array}{ccccccccccc} 0 \to H^1(T/R,U) & \to & H^1(S/R,U) & \to & H^1(S/T,U) & \to & H^2(T/R,U) & \to & H^2(S/R,U) & \to & \cdots \\ \uparrow & & \uparrow & & \uparrow & & \uparrow & & \uparrow & & \\ 0 \to H^1(R,G_1,U) & \to & H^1(R,G,U) & \to & H^1(R,G/G_1,U) & \to & H^2(R,G_1,U) & \to & H^2(R,G,U) & \to & \cdots \end{array}$$

To prove theorem 2, follow the arguments used to prove Proposition 4.1 and Theorem 4.3 in [5]. The exactness of the first row of the diagram is obtained from a spectral sequence associated with a bicomplex having terms $U(T^{m+1} \otimes S^{n+1})$. But letting $J = R[G]$ and $J_1 = R[G_1]$,

$$T^{m+1} \otimes S^{n+1} \approx J_1^m \otimes T \otimes S^{n+1} \approx J_1^m \otimes J_1 \otimes J^n \otimes S = J_1^{m+1} \otimes J^n \otimes S.$$

Consequently, there is a sub-bicomplex with terms $U(J_1^{m+1} \otimes J^n)$; and by tracing the arguments used to derive the first row of the above diagram, the second row is obtained from this sub-bicomplex.

Remarks: For $n \geq 1$, the homomorphism from $H^n(R,G/G_1,U)$ into $H^n(S/T,U)$ is the composition of the natural homomorphism of $H^n(R,G/G_1,U)$ into $H^n(T,G/G_1,U)$, which can be shown to be surjective, and the canonical homomorphism of $H^n(T,G/G_1,U)$ into $H^n(S/T,U)$. Under the hypotheses of theorem 2, $H^2(S/R,U) \approx B(S/R)$ according to [5, Lemma 3]. Finally if R is a field and S is a finite dimensional, modular, purely inseparable field extension of R, the surjectivity of the homomorphism of $H^2(R,G,U)$ into $H^2(S/R,U)$ has been proved by Hoechsmann [4].

References

1. S. U. Chase and A. Rosenberg, Amitsur cohomology and the Brauer group, Memoirs Amer. Math. Soc. No. 52(1965).

2. ___________, A theorem of Harrison, Kummer theory and Galois algebras, Nagoya Math. J. vol. 27(1966) pp. 663-685.

3. S. U. Chase and M. E. Sweedler, Hopf algebras and Galois theory, Lecture Notes in Math. No. 97, Springer Verlag, Berlin, 1969.

4. K. Hoechsmann, Algebras split by a given purely inseparable field, Proc. Amer. Math. Soc. 14(1963) pp. 768-776.

5. A. Rosenberg and D. Zelinsky, Amitsur's complex for inseparable fields, Osaka Math. J. vol. 14(1962) pp. 219-240.

6. S. Yuan, Brauer groups for inseparable fields, Amer. J. Math. vol. 96(1974) pp. 430-447.

7. D. Zelinsky, Berkson's theorem, Israel J. Math. vol. 2(1964) pp. 205-209.

NON-ADDITIVE RING AND MODULE THEORY IV

The Brauer Group of a Symmetric Monoidal Category

Bodo Pareigis

In [5],[6] and [7] we introduced general techniques in the theory of a monoidal category, i.e. of a category $\mathcal{C}$ with a bifunctor $\otimes: \mathcal{C} \times \mathcal{C} \longrightarrow \mathcal{C}$, an object $I \in \mathcal{C}$ and natural isomorphisms $\alpha: A \otimes (B \otimes C) \cong (A \otimes B) \otimes C$, $\lambda: I \otimes A \cong A$ and $\rho: A \otimes I \cong A$ which are coherent in the sense of [3, VII. 2]. In this paper we want to introduce the notion of a Brauer group of $\mathcal{C}$. For this purpose we are going to assume that $\mathcal{C}$ is symmetric, i.e. that there is a natural isomorphism $\gamma: A \otimes B \cong B \otimes A$ which is coherent with α, λ and ρ [3]. One of the main models for such a category $\mathcal{C}$ is, apart from the category of k-modules for a commutative ring k, the dual of the category of C-comodules for a cocommutative coalgebra C. This category is a symmetric monoidal category, but it is not closed.

Another type of monoidal categories, which are not symmetric but which allow the construction of Brauer groups, are for example categories of dimodules over a commutative, cocommutative Hopf algebra [2]. Their general theory will be discussed elsewhere.

In many special cases of symmetric monoidal categories the basic object I turns out to be projective, i.e. the functor $\mathcal{C}(I,-)$ preserves epimorphisms. In the general situation, however, it turns out that there may be constructed two Brauer groups $\mathcal{B}_1(\mathcal{C})$ and $\mathcal{B}_2(\mathcal{C})$ and a group-homomorphism $\mathcal{B}_2(\mathcal{C}) \longrightarrow \mathcal{B}_1(\mathcal{C})$, which is an isomorphism if I projective.

We will construct these two Brauer groups and discuss under which condition for a functor $F: \mathcal{C} \longrightarrow \mathcal{D}$ we get an induced homomorphism $\mathcal{B}_i(F): \mathcal{B}_i(\mathcal{C}) \longrightarrow \mathcal{B}_i(\mathcal{D})$.

<u>Preliminaries</u>

In [7] we proved analogues of the Morita Theorems which will be used in this paper. For the convenience of the reader we will collect the most important definitions and facts of [5],[6] and [7].

If P is an object of C we denote by $P(X)$ the set $C(X,P)$ for $X \in C$. Elements in $P \boxtimes Q\ (X)$ will often be denoted by $p \boxtimes q$. If the functor $C(P \boxtimes -,Q)$ is representable then the representing object is $[P,Q]$, so that $C(P \boxtimes X,Q) \cong C(X,[P,Q]) = [P,Q](X)$. The "evaluation" $P(X) \times [P,Q](Y) \longrightarrow Q(X \boxtimes Y)$, induced by the composition of morphisms, is denoted by $P(X) \times [P,Q](Y) \ni (p,f) \longmapsto \langle p\rangle f \in Q(X \boxtimes Y)$. Thus the "inner morphism sets" $[P,Q]$ operate on P from the right. In [5, Proposition 3.2] we prove that any natural transformation $P(X) \longrightarrow Q(X \boxtimes Y)$, natural in X, is induced by a uniquely determined element of $[P,Q](Y)$, if $[P,Q]$ exists.

We call an object $P \in C$ finite or finitely generated projective if $[P,I]$ and $[P,P]$ exist and if the morphism $[P,I] \boxtimes P \longrightarrow [P,P]$ induced by $P(X) \times [P,I](Y) \times P(Z) \ni (p,f,p') \longmapsto \langle p\rangle fp' \in P(X \boxtimes Y \boxtimes Z)$ is an isomorphism. For $[P,I] \boxtimes P \longrightarrow [P,P]$ to be an isomorphism it is necessary and sufficient that there is a "dual basis" $f_o \boxtimes p_o \in [P,I] \boxtimes P(I)$ such that $\langle p\rangle\, f_o p_o = p$ for all $p \in P(X)$ and all $X \in C$. The difference between finite and finitely generated projective objects, as discussed in [8], does not appear here.

A finite object P is called faithfully projective if the morphism $P \underset{[P,P]}{\boxtimes} [P,I] \longrightarrow I$, induced by the evaluation, is an isomorphism. This is equivalent to the existence of $p_1 \underset{[P,P]}{\boxtimes} f_1 \in P \underset{[P,P]}{\boxtimes} [P,I]\ (I)$ with $\langle p_1\rangle f_1 = 1 \in I(I)$. If there exists an element $p_1 \boxtimes f_1 \in P \boxtimes [P,I]\ (I)$ with $\langle p_1\rangle f_1 = 1$, then P is called a progenerator. Now $P \boxtimes [P,I] \longrightarrow P \underset{[P,P]}{\boxtimes} [P,I]$ is an epimorphism; if I is projective, then P is faithfully projective iff P is a progenerator.

Let ${}_AC$ denote the category of A-objects in C with A a monoid.

Then a functor ${}_A\mathcal{C} \ni X \longmapsto P \boxtimes_A X \in {}_B\mathcal{C}$ with ${}_BP_A$ a left B right A biobject is a category equivalence iff ${}_BP$ is faithfully projective and $A \cong {}_B[P,P]$ as has been proved in [7]. For this Morita equivalence all the usual conclusions hold, in particular the centers ${}_A[A,A]_A$ and ${}_B[B,B]_B$ of A resp. B are isomorphic monoids if they exist.

<u>The Brauer group $\mathcal{B}_1(\mathcal{C})$.</u>

Let $\mathcal{C}$ be a symmetric monoidal category. A monoid A in $\mathcal{C}$ is called 1-Azumaya if $\mathcal{C} \ni X \longmapsto A \boxtimes X \in {}_A\mathcal{C}_A$ is an equivalence of cate - gories. Thus the Morita Theorems, in particular [7, Theorem 5.1] can be applied.

<u>Proposition</u> 1: <u>A monoid</u> A <u>is</u> 1-<u>Azumaya iff</u> $A \in \mathcal{C}$ <u>is faithfully projective and</u>

$$\psi: A \boxtimes A\,(X) \ni a \boxtimes b \longmapsto (A(Y) \ni c \longmapsto acb \in A(X \boxtimes Y)) \in [A,A](X)$$

<u>is an isomorphism</u>.

Proof: Let A be faithfully projective and ψ be an isomorphism. Let A^{op} be the monoid on A with inverse multiplication. Then $\psi: A^{op} \boxtimes A \longrightarrow [A,A]$ is an isomorphism of monoids. Thus the categories ${}_A\mathcal{C}_A \cong \mathcal{C}_{[A,A]}$ are equivalent with the functor $M \longmapsto M$. Furthermore $\mathcal{C} \ni X \longmapsto A \boxtimes X \in \mathcal{C}_{[A,A]}$ is an equivalence by [7, Theorem 5.4] hence $\mathcal{C} \ni X \longmapsto A \boxtimes X \in {}_A\mathcal{C}_A$ is an equivalence and A is 1-Azumaya.

Conversely if A is 1-Azumaya the morphism ψ which exists for any monoid A induces a commutative diagram

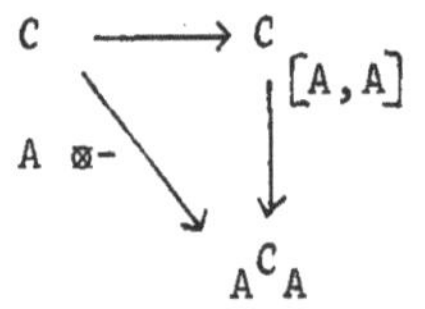

Hence $A \in C$ is faithfully projective in C by [7, Theorem 5.1] and so ψ must induce a category isomorphism and must even be an isomorphism [7, Theorem 5.1 d)] . Q.E.D.

Recall that the center of a monoid A is the object ${}_A[A,A]_A$, if it exists. ${}_A[A,A]_A$ is the representing object of the functor $C \in X \longmapsto {}_AC_A(A \boxtimes X, A) \in S$. Observe that for this definition we need the symmetry of the monoidal category C .

Let A be a monoid in C . Since we have ${}_A[A,A]_A(X) \subseteq A(X)$, the inclusion given by the isomorphism ${}_A[A,A] \cong A$ using the multiplication with A from the right, it is easy to see that

$a \in {}_A[A,A]_A(X)$ iff $ab = ba$ for all $b \in A(Y)$ and all $Y \in C$. Since $ab = ba$ for all $a \in \mathrm{Im}(\eta(X): I(X) \longrightarrow A(X))$ and all $b \in A(Y)$, all $X,Y \in C$, we get that $\eta(X)$ maps $I(X)$ into ${}_A[A,A]_A(X)$. If this morphism is an isomorphism then A is called a central monoid.

Let A be 1-Azumaya. Then $[A,A]$ is Morita equivalent to I hence the center I of I coincides with the center of $[A,A]$ via the morphism $\eta: I \longrightarrow [A,A]$ [7, Corollary 6.3] . Thus the morphism

$$I \xrightarrow{\eta} A \xrightarrow{\varphi} A^{op} \boxtimes A \xrightarrow{\psi} [A,A]$$

is injective. φ is defined by $\varphi(a) = 1 \boxtimes a$. Hence $I(X)$ is contained in the center ${}_A[A,A]_A(X)$. Now let $a \in A(X)$ such that $ab = ba$ for all $b \in A(Y)$, all $Y \in C$. Then $(1 \boxtimes a)(b \boxtimes c) = b \boxtimes ac = b \boxtimes ca = (b \boxtimes c)(1 \boxtimes a)$ for all $b \boxtimes c \in A^{op} \boxtimes A(Y)$. Thus $\varphi(a)$ is in the center of $A^{op} \boxtimes A$ or $\psi\varphi(a)$ in the center of $[A,A]$, which was I . Furthermore φ is a monomorphism, even a section with retract $\mu: A \boxtimes A \longrightarrow A$. Thus $a \in \mathrm{Im}(\eta(X): I(X) \longrightarrow A(X))$, so that I is the center of A .

<u>Corollary 2</u>: <u>If</u> A <u>is</u> 1-<u>Azumaya</u> <u>then</u> A <u>is</u> <u>central</u>.

<u>Proposition 3</u>: <u>Let</u> A , B <u>be</u> 1-<u>Azumaya</u> <u>then</u> $A \boxtimes B$ <u>is</u> 1-<u>Azumaya</u>.

Proof: Let $f_o \otimes a_o$ resp. $g_o \otimes b_o$ be a dual basis for A resp. B. Then $f_o \otimes g_o \otimes a_o \otimes b_o \in [A \otimes B, I] \otimes A \otimes B(I)$ is a dual basis for $A \otimes B$ where we identified $[A,I] \otimes [B,I]$ with $[A \otimes B, I]$. Furthermore we have $[A,A] \otimes [B,B] \cong [A \otimes [A,I], [B,B]] \cong [A \otimes B \otimes [A,I], B] \cong [A \otimes B, A \otimes B]$ since A and B are finite [8, Theorem 1.2]. Hence $[A \otimes B, I]$ and $[A \otimes B, A \otimes B]$ exist.

Let $A^e = A^{op} \otimes A$ and $A' \in {}_{A^e}C_I$ be the dual $[A,I]$ of A. With the analogous notation for B we get

$$(A \otimes B) \otimes_{A^e \otimes B^e} (A' \otimes B') \cong (A \otimes_{A^e} A') \otimes (B \otimes_{B^e} B') \cong I$$

since A, B are faithfully projective. Hence $A \otimes B$ is faithfully projective in C.

Finally since $[A,A] \otimes [B,B] \cong [A \otimes B, A \otimes B]$ we get that $\psi: A \otimes B \otimes A \otimes B \longrightarrow [A \otimes B, A \otimes B]$ is an isomorphism.

<u>Proposition</u> 4: <u>Let</u> P <u>be faithfully projective then</u> $[P,P]$ <u>is</u> 1-<u>Azumaya</u>.

Proof: We know that $[P,I] \otimes P \cong [P,P]$ as $[P,P]$ - $[P,P]$ - objects. Furthermore $C \ni X \longmapsto [P,I] \otimes X \in {}_{[P,P]}C$ and $C \ni X \longmapsto X \otimes P \in C_{[P,P]}$ are equivalences. Hence $C \ni X \longmapsto [P,P] \otimes X \cong [P,I] \otimes X \otimes P \in {}_{[P,P]}C_{[P,P]}$ is an equivalence, since ${}_{[P,P]}C \ni Y \longmapsto Y \otimes P \in {}_{[P,P]}C_{[P,P]}$ is also an equivalence.

<u>Proposition</u> 5: <u>Let</u> P, Q <u>be faithfully projective, then</u> $P \otimes Q$ <u>is faithfully projective and</u> $[P,P] \otimes [Q,Q] \cong [P \otimes Q, P \otimes Q]$ <u>as monoids</u>.

Proof: Since P and Q are finite we get for all $X, Y \in C$ that $\varphi: [P,X] \otimes [Q,Y] \ni f \otimes g \longmapsto (p \otimes q \longmapsto \langle p \rangle f \otimes \langle q \rangle g) \in [P \otimes Q, X \otimes Y]$ is an isomorphism and the right side exists. In particular

$[P \otimes Q, I]$ and $[P \otimes Q, P \otimes Q]$ exist. Furthermore we have $\langle p \otimes q\rangle\varphi(f \otimes g)\varphi(f' \otimes g') = \langle p\rangle ff' \otimes \langle q\rangle gg' = \langle p \otimes q\rangle\varphi(ff' \otimes gg')$ and $\varphi(id_P \otimes id_Q) = id_{P \otimes Q}$, hence $\varphi: [P,P] \otimes [Q,Q] \longrightarrow [P \otimes Q, P \otimes Q]$ is an monoid isomorphism.

If $f_o \otimes p_o$ resp. $g_o \otimes q_o$ are dual bases of P and Q then $\varphi(f_o \otimes g_o) \otimes (p_o \otimes q_o)$ is a dual basis of $P \otimes Q$ for $\langle p \otimes q\rangle\varphi(f_o \otimes g_o)(p_o \otimes q_o) = \langle p\rangle f_o p_o \otimes \langle q\rangle g_o q_o = p \otimes q$. Hence $P \otimes Q$ is finite.

Now let $p_1 \underset{[P,P]}{\otimes} f_1$ resp. $q_1 \underset{[Q,Q]}{\otimes} g_1$ be elements such that $\langle p_1\rangle f_1 = 1$ and $\langle q_1\rangle g_1 = 1$. Then $(p_1 \otimes q_1) \underset{B}{\otimes} \varphi(f_1 \otimes g_1)$ with $B = [P \otimes Q, P \otimes Q]$ has the property $\langle p_1 \otimes q_1\rangle \varphi(f_1 \otimes g_1) = \langle p_1\rangle f_1 \langle q_1\rangle g_1 = 1$. Thus $P \otimes Q$ is faithfully projective.

Now we can define the Brauer group $\mathcal{B}_1(\mathcal{C})$ of a symmetric monoidal category $\mathcal{C}$. Let $\mathcal{A}$ be the(illegitimate) set of isomorphism classes of 1-Azumaya monoids A in $\mathcal{C}$. Then we define an equivalence relation on $\mathcal{A}$ by $\bar{A} \sim \bar{B}$ iff there exist faithfully projective $P,Q \in \mathcal{C}$ such that $A \otimes [P,P] \cong B \otimes [Q,Q]$ as monoids. Denote the set of equivalence classes by $\mathcal{B}_1(\mathcal{C})$. $\mathcal{B}_1(\mathcal{C})$ becomes a commutative group in the usual way by $[A][B] = [A \otimes B]$ with unit element $[I]$ and inverse $[A^{op}]$ for $[A]$, where A^{op} is the 1-Azumaya monoid A with inverse multiplication $A \otimes A \xrightarrow{\gamma} A \otimes A \xrightarrow{\mu} A$.

<u>Separable monoids</u>

Let A be a monoid in $\mathcal{C}$. A is called a separable monoid if the multiplication $\mu: A \otimes A \longrightarrow A$ has a splitting $\sigma: A \longrightarrow A \otimes A$ in ${}_A\mathcal{C}_A$ such that $\mu\sigma = id_A$. Observe that $A(X) \ni a \longmapsto a \otimes 1 \in A \otimes A(X)$ is a splitting for μ in ${}_A\mathcal{C}$ but it is no A-right-morphism.

__Proposition__ 6: __Let__ $A \in C$ __be a monoid__. __Equivalent are__

a) A __is separable__

b) __There is an element__ $a \boxtimes b \in A \boxtimes A(I)$ __such that__

i) $\forall\, c \in A(X)$: $ca \boxtimes b = a \boxtimes bc \in A \boxtimes A\,(X)$,

ii) $ab = 1 \in A(I)$.

Proof: a) $\Rightarrow$ b): Define $a \boxtimes b := \sigma(1)$. Then $1 = \mu\sigma(1) = \mu(a \boxtimes b) = ab$ which is condition (ii). Since σ is an A-A-morphism we have $ca \boxtimes b = c\sigma(1) = \sigma(c) = \sigma(1)c = a \boxtimes bc$ for all $a \in A(X)$.

b) $\Rightarrow$ a): Let $\sigma\colon A(X) \longrightarrow A \boxtimes A(X)$ be defined by $\sigma(c) = ca \boxtimes b$. By (i) σ is an A-A-morphism. By (ii) we get $\mu\sigma(c) = \mu(ca \boxtimes b) = cab = c$, hence $\mu\sigma = id_A$.

Observe that (i) does not depend on a symmetry in C , since $a \boxtimes b \in A \boxtimes A(I)$ and $I \boxtimes X \cong X \boxtimes I$ even without a symmetry. The element $a \boxtimes b$ will be called a Casimir element.

Every Casimir element $a \boxtimes b \in A \boxtimes A(I)$ induces a map $Tr\colon C\,(M,N) \ni f \longmapsto (M(X) \ni m \longmapsto af(bm) \in N(X)) \in {}_AC\,(M,N)$ for any two objects $M,N \in {}_AC$. In fact for any $c \in A(Y)$ we have $ca \boxtimes b = a \boxtimes bc$ hence $c(af(bm)) = (ca)f(bm) = af(bcm) = af(b(cm))$. This map is called the trace map.

Since the trace map is a natural transformation, natural in X ,

$$Tr\colon C(M \boxtimes X,N) \longrightarrow {}_AC(M \boxtimes X,N) ,$$

we get $Tr\colon [M,N] \longrightarrow {}_A[M,N]$, if both objects exist.

Since $ab = 1$ we even get that

$${}_AC(M,N) \longrightarrow C(M,N) \xrightarrow{Tr} {}_AC(M,N)$$

is the identity on ${}_AC(M,N)$ since $Tr(f)(m) = af(bm) = abf(m) = f(m)$ and hence $Tr(f) = f$, if $f \in {}_AC(M,N)$. Similarly ${}_A[M,N] \longrightarrow [M,N] \longrightarrow {}_A[M,N]$ is the identity on ${}_A[M,N]$.

If $M,N \in {}_AC_A$ then we clearly get $Tr\colon C_A(M,N) \longrightarrow {}_AC_A(M,N)$ and ${}_AC_A(M,N) \longrightarrow C_A(M,N) \xrightarrow{Tr} {}_AC_A(M,N)$ is the identity on ${}_AC_A(M,N)$. The same holds for $[M,N]_A$ and $_A[M,N]_A$.

If $_A[A,A]_A$ exists then

$$_A[A,A]_A \longrightarrow [A,A]_A \xrightarrow{Tr} {}_A[A,A]_A$$

is the identity on $_A[A,A]_A$. Observe that $[A,A]_A$ exists, since $[A,A]_A \cong A$. Since the last isomorphism is an antiisomorphism of monoids, $_A[A,A]_A$ is the center of A and $_A[A,A]_A \longrightarrow A$ is a monoid homomorphism, we get

Proposition 7 [1, Prop. 1.2]: *If A is a separable monoid, then the center $_A[A,A]_A$ (if it exists) is a "direct summand" of A .*

Let $f: A \longrightarrow B$ be monoid homomorphism. $P \in {}_BC$ is called (B,A)-projective if for each commutative diagram

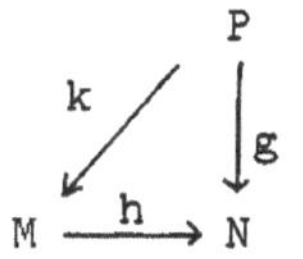

with g,h in $_BC$ and k in $_AC$ there is a $g' \in_B C(P,M)$ with $hg' = g$. The dual notion is that of a (B,A)-injective object [8] .

Proposition 8: *Let A be a separable monoid. Then every A-object is (A,I)-projective and (A,I)-injective.*

Proof: Let $g \in {}_AC(P,N)$, $h \in {}_AC(M,N)$ and $k \in C(P,M)$ be given such that $hk = g$. Then $g = Tr(g) = Tr(hk) = h\,Tr(k)$ and $Tr(k) \in {}_AC(P,M)$, so that P is (A,I)-projective. Just by reversing the arrows one can prove that each object in $_AC$ is (A,I)-injective.

In [8] we prove that in $(C,\times,E)$, a monoidal category with the product as tensor-product and E a final object, there are no non-trivial finite objects. In Theorem 14 we shall show that $[P,P]$ is a separable monoid for certain finite objects $P \in C$. So this con - struction will not produce examples of separable monoids in $(C, \times, E)$. In fact, there are no non-trivial separable monoids in C at all.

Proposition 9: *Let A be a separable monoid in the monoidal category $(C, \times, E)$. Then $A \cong E$ as monoids.*

Proof: Let (a,b) be the Casimir element for A . Then $(ca,b) = (a,bc)$ for all $c \in A(X)$, hence $ca = a$ and $b = bc$. Here we use $A \times A(X) = A(X) \times A(X)$ and $A(E) \longrightarrow A(X)$ by the unique mor - phism $X \longrightarrow E$. We also have $1c = c$ and $ab = 1$, hence $c = 1c = abc = ab = 1$ for all $c \in A(X)$, so that $A(X) = \{1\}$ which proves $A \cong E$ .Finally observe that E has a unique monoid structure.

If A and B are monoids in C , then $A \boxtimes B$ is a monoid by $(a_1 \boxtimes b_1) \cdot (a_2 \boxtimes b_2) = a_1a_2 \boxtimes b_1b_2$.

Proposition 10: *Let A and B be separable monoids. Then $A \boxtimes B$ is separable.*

Proof: Let $a_1 \boxtimes a_2$ and $b_1 \boxtimes b_2$ be Casimir elements of A resp. B. Then $(a_1 \boxtimes b_1) \boxtimes (a_2 \boxtimes b_2)$ is a Casimir element for $A \boxtimes B$. In fact let $x \boxtimes y \in A \boxtimes B(X)$ then

$$(x \boxtimes y)(a_1 \boxtimes b_1) \boxtimes (a_2 \boxtimes b_2) = (xa_1 \boxtimes yb_1) \boxtimes (a_2 \boxtimes b_2) =$$
$$(a_1 \boxtimes b_1) \boxtimes (a_2x \boxtimes b_2y) = (a_1 \boxtimes b_1) \boxtimes (a_2 \boxtimes b_2)(x \boxtimes y) .$$

Furthermore $(a_1 \boxtimes b_1)(a_2 \boxtimes b_2) = a_1a_2 \boxtimes b_1b_2 = 1 \boxtimes 1$.

Proposition 11: *Let A be a separable monoid with Casimir element $a \boxtimes b$. Assume that ${}_A[A,A]_A$ exists and that $I \longrightarrow A$ is a mono - morphism. Then A is central if and only if $axb \in I(X)$ for all $x \in A(X)$.*

Proof: Since $ca \otimes b = a \otimes bc$ for all $c \in A(Y)$ we get $c(axb) = (axb)c$ hence $axb \in {}_A[A,A]_A(X)$ for all $x \in A(X)$. If A is central then $axb \in {}_A[A,A]_A(X) = I(X)$ for all $x \in A(X)$. Conversely let $axb \in I(X)$ for all $x \in A(X)$. Let $x \in {}_A[A,A]_A(X)$ then $x = xab = axb \in I(X)$ hence ${}_A[A,A]_A(X) = I(X)$.

The Brauer group $\mathcal{B}_2(C)$

A monoid A is called 2-Azumaya if $[A,I]$ and $[A,A]$ exist and $\eta: I \longrightarrow A$ is a monomorphism and if there are elements $a \otimes b \in A \otimes A(I)$ and $c \otimes d \otimes e \in A \otimes A \otimes A(I)$ such that

i) $\forall X \in C \quad \forall x \in A(X): xa \otimes b = a \otimes bx$,

ii) $ab = 1 \in A(I)$,

iii) $ac \otimes dbe = 1 \otimes 1 \in A \otimes A\,(I)$,

iv) $\forall X \in C \quad \forall x \in A(X): axb \in I(X)$.

Clearly a 2-Azumaya monoid is a central, separable monoid which follows from i), ii), and iv) by Proposition 6 and 11 . We do not know if the existence of $c \otimes d \otimes e$ with iii) follows from the other conditions.

Theorem 12: Let A be a monoid in C . Equivalent are

a) A is 2-Azumaya.

b) $A \in C$ is a progenerator and the morphism

$\psi: A \otimes A(X) \ni x \otimes y \longmapsto (A(Y)) \ni z \longmapsto xzy \in A(X \otimes Y)) \in [A,A](X)$

is an isomorphism.

c) A is separable and 1-Azumaya.

Proof: Let A be 2-Azumaya. Define $\varphi: A \otimes A \longrightarrow [A,I]$ by $\langle z\rangle(\varphi(x \otimes y)) := axzyb$ where $a \otimes b$ is a Casimir element for A . Then $axzyb \in I(X)$ by iv) hence φ is well-defined. Now

$\varphi(c \otimes d) \otimes e$ is a dual basis for A since $\langle x\rangle(\varphi(c \otimes d)e) = acxdbe = x$ for all $x \in A(X)$. So A is finite.

Now we show that $A \otimes [A,I] \longrightarrow I$, the morphism induced by the evaluation, is rationally surjective, i.e. that $A \otimes [A,I]\,(I) \longrightarrow I(I)$ is surjective. We have to find $a_1 \otimes f_1 \in A \otimes [A,I](I)$ with $\langle a_1\rangle f_1 = 1 \in I(I)$. Take $a_1 = 1$ and $f_1 = (Tr\colon A \longrightarrow {}_A[A,A]_A \cong I)$, the last isomorphism exists in view of Propositions 6 and 11 by the properties i), ii), and iv) of A. Then $\langle 1\rangle f_1 = 1$, hence A is a progenerator.

To show that ψ is an isomorphism we construct the inverse morphism

$$[A,A](X) \ni \sigma \longmapsto \langle ac\rangle\sigma db \otimes e \in A \otimes A(X)\ .$$

This morphism is in fact an inverse of ψ since

$$\langle x\rangle\ \psi\,(\langle ac\rangle\sigma db \otimes e) = \langle ac\rangle\sigma dbxe = \langle xac\rangle\sigma dbe$$
$$= \langle x1\rangle\sigma 1 = \langle x\rangle\sigma\ ,$$

hence $\psi\,(\langle ac\rangle\sigma db \otimes e) = \sigma$, and

$$xacydb \otimes e = x \otimes acydbe = x \otimes 1y1 = x \otimes y\ .$$

Now assume that b) holds. By Proposition 1 the monoid A is 1-Azumaya. Let $f_o \otimes a_o$ be a dual basis for A and $a_1 \otimes f_1 \in A \otimes [A,I](I)$ such that $\langle a_1\rangle f_1 = 1 \in I(I)$. Let $g_o \in [A,I](I)$ be defined by $\langle x\rangle g_o = \langle xa_1\rangle f_1$. Let $a \otimes b \in A \otimes A(I)$ be the element which corresponds to $g_o \otimes 1 \in [A,I] \otimes A(I)$ under the isomorphism $A \otimes A \cong [A,A] \cong [A,I] \otimes A$. Then we have $ab = a1b = \langle 1\rangle g_o 1 = \langle a_1\rangle f_1 1 = 1 \in A(I)$. Furthermore we have $xayb = x\langle y\rangle g_o = \langle y\rangle g_o x = aybx$ for all $y \in A(Y)$, hence $xa \otimes b = a \otimes bx$ for all $x \in A(X)$. So $a \otimes b$ is a Casimir element for A. Thus c) holds.

Assume that c) holds. By Proposition 1 A is faithfully projective and ψ is an isomorphism. Construct $a_1 \otimes f_1 \in A \otimes [A,I](I)$ with $\langle a_1\rangle f_1 = 1 \in I(I)$ as in part one of the proof. Then b) holds.

We still have to show that b) and c) imply a). Let $f_o^1 \otimes a_o^1$ and

$f_o^2 \otimes a_o^2$ be two copies of the dual basis of A. Let $a_1 \otimes f_1 \in A \otimes [A,I](I)$ with $<a_1>f_1 = 1$ be given. Then define $a \otimes b$ as above corresponding to g_o and $c \otimes d \otimes e :=$ $u \otimes v \otimes xy \in A \otimes A \otimes A\ (I)$, where $u \otimes v \otimes x \otimes y \in A \otimes A \otimes A \otimes A\ (I)$ corresponds to $f_o^1 \otimes a_o^2 \otimes f_o^2 \otimes a_o^1 \in [A,I] \otimes A \otimes [A,I] \otimes A(I)$ under the isomorphisms

$[A,I] \otimes A \otimes [A,I] \otimes A \cong [A,A] \otimes [A,A] \cong A \otimes A \otimes A \otimes A$. Then $aczdbe = auzvbxy = <uzv>g_o xy = <<<z>f_o^1 a_o^2 a_1>f_1 1>f_o^2 a_o^1 =$ $<z>f_o^1 <a_o^2 a_1> f_1 <1>f_o^2 a_o^1 = <<1>f_o^2 a_o^2 a_1>f_1 <z>f_o^1 a_o^1 = z = 1z1$ for all $z \in A(Z)$, hence $ac \otimes dbe = 1 \otimes 1$. Thus iii) for a monoid to be 2-Azumaya holds. i) and ii) hold by Proposition 1, iv) by Proposition 11.

Corollary 13: Let A and B be 2-Azumaya, then $A \otimes B$ is 2-Azumaya.

Proof: In view of the equivalence of a) and c) in Theorem 12 this follows from Proposition 3 and Proposition 10.

Theorem 14: Let $P \in C$ be a progenerator. Then $[P,P]$ is 2-Azumaya.

Proof: By Proposition 4 we get that $[P,P]$ is 1-Azumaya so that we only have to show that $[P,P]$ is separable. Let $f_o \otimes p_o$ be a dual basis for P and $p_1 \otimes f_1 \in P \otimes [P,I](I)$ with $<p_1>f_1 = 1$. Identify $[P,P]$ with $[P,I] \otimes P$ with the multiplication $(f \otimes p)(f' \otimes p') = f \otimes <p>f'p'$. Then define $a \otimes b := (f_o \otimes p_1) \otimes (f_1 \otimes p_o)$. For every $g \otimes q \in [P,I] \otimes P(X)$ we have

$$(g \otimes q)(f_o \otimes p_1) \otimes (f_1 \otimes p_o) = (g \otimes <q>f_o p_1) \otimes (f_1 \otimes p_o) =$$

$$(g \otimes p_1) \otimes (f_1 \otimes <q>f_o p_o) = (g \otimes p_1) \otimes (f_1 \otimes q) =$$

$(f_o\langle p_o\rangle g \otimes p_1) \otimes (f_1 \otimes q) = (f_o \otimes p_1) \otimes (f_1 \otimes \langle p_o\rangle gq) =$

$(f_o \otimes p_1) \otimes (f_1 \otimes p_o)(g \otimes q)$

so that b) i) of Proposition 6 holds. Furthermore

$(f_o \otimes p_1)(f_1 \otimes p_o) = f_o \otimes \langle p_1\rangle f_1 p_o = f_o \otimes p_o$,

which corresponds to $1 \in [P,P](I)$, shows b) ii) .

It may be interesting to have an explicit description of the element $c \otimes d \otimes e$ in the definition of 2-Azumaya for this case $[P,P]$. Let $f_o^i \otimes p_o^i$, i = 1, 2, 3 be copies of the dual basis of P . Then $c \otimes d \otimes e := (f_o^1 \otimes p_o^2) \otimes (f_o^3 \otimes p_o^1) \otimes (f_o^2 \otimes p_o^3)$ satisfies condition iii) for 2-Azumaya as is easily checked.

To define a Brauer group of 2-Azumaya monoids we need one more lemma.

<u>Lemma</u> 15: <u>Let</u> P <u>and</u> Q <u>be progenerators</u>. <u>Then</u> $P \otimes Q$ <u>is a progenerator and</u> $[P,P] \otimes [Q,Q] \cong [P \otimes Q, P \otimes Q]$ <u>as monoids</u>.

Proof: Let $p_1 \otimes f_1$ resp. $q_1 \otimes g_1$ with $\langle p_1\rangle f_1 = 1$ resp. $\langle q_1\rangle g_1 = 1$ be given. Then form the element $(p_1 \otimes q_1) \otimes \varphi(f_1 \otimes g_1) \in (P \otimes Q) \otimes [P \otimes Q, I](I)$, where $\varphi: [P,I] \otimes [Q,I] \cong [P \otimes Q, I]$ is the isomorphism used in the proof of Proposition 5 . We get $(p_1 \otimes q_1)\varphi(f_1 \otimes g_1) = \langle p_1\rangle f_1 \langle q_1\rangle g_1 = 1$, hence $P \otimes Q$ is a progenerator in view of Proposition 5 .

Now we can define the Brauer group $B_2(C)$, using 2-Azumaya monoids, in the same way as $B_1(C)$. Since each 2-Azumaya monoid is 1-Azumaya and since each progenerator is faithfully projective we get a group homomorphism $\xi: B_2(C) \longrightarrow B_1(C)$. Since the notions of progenerator and faithfully projective coincide, if $I \in C$ is projective, the notions of 1-Azumaya and 2-Azumaya coincide by Theorem 12, b) and Proposition 1 . So does the equivalence relation used in the construction of the two Brauer groups and we get

Theorem 16: The group homomorphism $\xi\colon B_2(C) \longrightarrow B_1(C)$ is the identity in case $I \in C$ is projective.

Splitting Azumaya monoids by monoidal functors.

Now we want to discuss the behaviour of the Brauer groups under a monoidal functor. Let C and D be symmetric monoidal categories and $F\colon C \longrightarrow D$ be a covariant functor. Denote the tensor products and the associativity, the symmetry and unity isomorphisms in C and D by the same signs $\otimes$, α, γ, λ, and ρ . Assume that there are natural transformations

$\delta\colon FX \otimes FY \longrightarrow F(X \otimes Y)$

$\zeta\colon J \longrightarrow FI$

such that the following diagrams commute

$$\begin{array}{ccc} FX \otimes FI & \xleftarrow{1 \otimes \zeta} & FX \otimes J \\ \downarrow{\scriptstyle\delta} & & \downarrow{\scriptstyle\rho} \\ F(X \otimes I) & \xrightarrow{F(\rho)} & FX \end{array}$$

$$\begin{array}{ccc} FI \otimes FX & \xleftarrow{\zeta \otimes 1} & J \otimes FX \\ \downarrow{\scriptstyle\delta} & & \downarrow{\scriptstyle\lambda} \\ F(I \otimes X) & \xrightarrow{F(\lambda)} & FX \end{array}$$

$$\begin{array}{ccccc} FX \otimes (FY \otimes FZ) & \xrightarrow{1 \otimes \delta} & FX \otimes F(Y \otimes Z) & \xrightarrow{\delta} & F(X \otimes (Y \otimes Z)) \\ \downarrow{\scriptstyle\alpha} & & & & \downarrow{\scriptstyle F(\alpha)} \\ (FX \otimes FY) \otimes FZ & \xrightarrow{\delta \otimes 1} & F(X \otimes Y) \otimes FZ & \xrightarrow{\delta} & F((X \otimes Y) \otimes Z) \end{array} .$$

If C and D are symmetric we require in addition the commutativity of

$$\begin{array}{ccc} FX \boxtimes FY & \xrightarrow{\gamma} & FY \boxtimes FX \\ \downarrow{\scriptstyle\delta} & & \downarrow{\scriptstyle\delta} \\ F(X \boxtimes Y) & \xrightarrow{F(\gamma)} & F(Y \boxtimes X) \end{array} .$$

Such a triple (F, δ, ζ) will be called a weakly monoidal functor.

Let $\pi\colon X \boxtimes [X,Y] \longrightarrow Y$ and $\tau\colon Y \longrightarrow [X, X \boxtimes Y]$ be front and back adjunction for the adjoint pair of functors $X \boxtimes -$ and $[X,-]$, if $[X,-]$ exists. Again we use the same notation in both categories $\mathcal{C}$ and $\mathcal{D}$. Let $\chi\colon \mathcal{C}(X \boxtimes Y, Z) \cong \mathcal{C}(Y, [X,Z])$ and $\omega\colon \mathcal{C}(Y, [X,Z]) \cong \mathcal{C}(X \boxtimes Y, Z)$ be the corresponding adjointness isomor - phisms in $\mathcal{C}$ resp. also in $\mathcal{D}$. It is an easy exercise in diagram chasing for adjoint functors to show that there is a natural trans - formation $\Phi\colon F[X,Y] \longrightarrow [FX,FY]$ whenever $[X,Y]$ and $[FX,FY]$ exist, just take $\Phi = \chi(F(\pi)\delta)$. Furthermore the diagrams

$$\begin{array}{ccc} FX \boxtimes F[X,Y] & \xrightarrow{\delta} & F(X \boxtimes [X,Y]) \\ \downarrow{\scriptstyle 1 \boxtimes \Phi} & & \downarrow{\scriptstyle F(\pi)} \\ FX \boxtimes [FX,FY] & \xrightarrow{\pi} & FY \end{array} \qquad \begin{array}{ccc} FY & \xrightarrow{F(\tau)} & F[X, X \boxtimes Y] \\ \downarrow{\scriptstyle\tau} & & \downarrow{\scriptstyle\Phi} \\ [FX, FX \boxtimes FY] & \xrightarrow{[1,\delta]} & [FX, F(X \boxtimes Y)] \end{array} \text{ and}$$

$$\begin{array}{ccc} J & \xrightarrow{j} & [FX,FX] \\ \downarrow{\scriptstyle\zeta} & & \downarrow{\scriptstyle\Phi} \\ FI & \xrightarrow{F(i)} & F[X,X] \end{array}$$

commute. Here $i\colon I \longrightarrow [P,P]$ is $\chi(\rho)$ where $\rho : X \boxtimes I \longrightarrow X$ and j is defined analogously in $\mathcal{D}$.

Omitting special arrows for the associativity α we get the commutative diagram on the next page.

If we abbreviate $[1, \pi \boxtimes 1]\tau$ by $\varphi : [X,Y] \boxtimes Z \longrightarrow [X, Y \boxtimes Z]$ then the following diagram (the outer frame of the given diagram) commutes

$$\begin{array}{ccccc} F[X,Y] \boxtimes FZ & \xrightarrow{\delta} & F([X,Y] \boxtimes Z) & \xrightarrow{F(\varphi)} & F([X, Y \boxtimes Z]) \\ \downarrow{\scriptstyle \Phi \boxtimes 1} & & & & \downarrow{\scriptstyle\Phi} \\ [FX,FY] \boxtimes FZ & \xrightarrow{\varphi} & [FX, FY \boxtimes FZ] & \xrightarrow{[1,\delta]} & [FX, F(Y \boxtimes Z)] \end{array} .$$

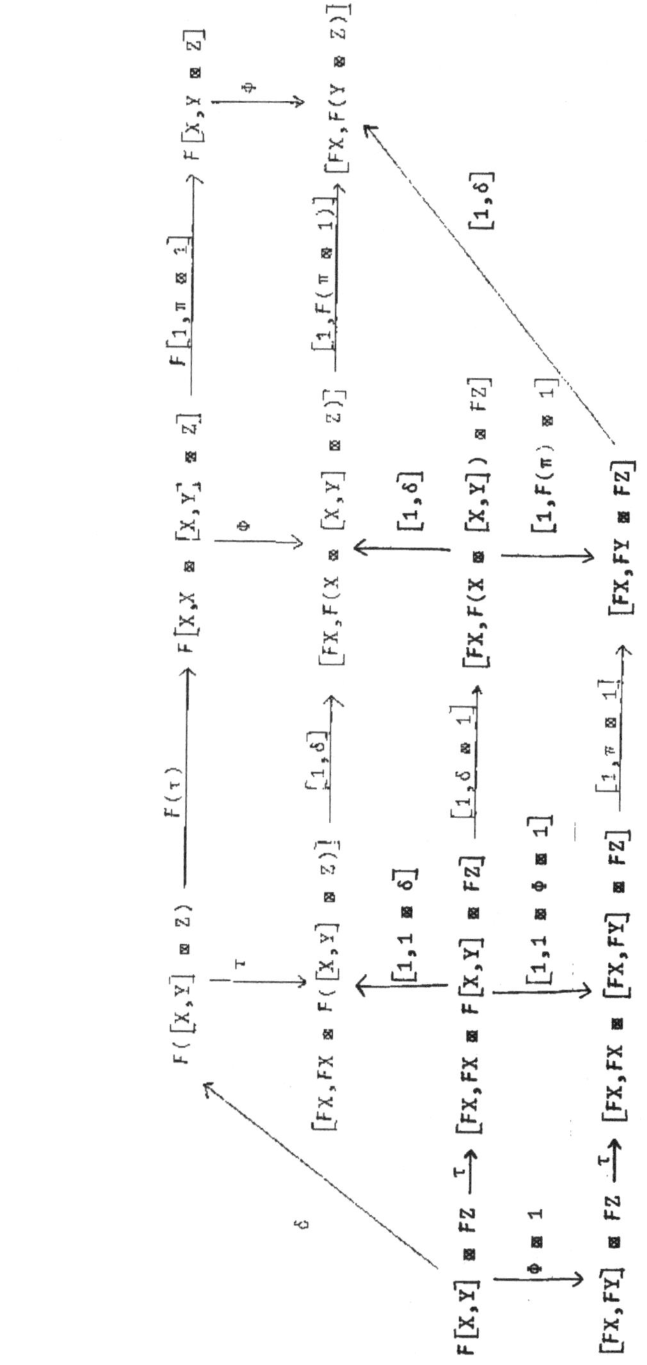
F([X,Y] ⊠ Z)
F(τ)
F[X,X ⊠ [X,Y] ⊠ Z]
F[1,π ⊠ 1]
F[X,Y ⊠ Z]
τ
Φ
Φ
[FX,FX ⊠ F([X,Y] ⊠ Z)]
[1,δ]
[FX,F(X ⊠ [X,Y] ⊠ Z)]
[1,F(π ⊠ 1)]
[FX,F(Y ⊠ Z)]
[1,1 ⊠ δ]
[1,δ]
δ
[1,δ]
F[X,Y] ⊠ FZ
τ
[FX,FX ⊠ F[X,Y] ⊠ FZ]
[1,δ ⊠ 1]
[FX,F(X ⊠ [X,Y]) ⊠ FZ]
Φ ⊠ 1
[1,1 ⊠ Φ ⊠ 1]
[1,F(π) ⊠ 1]
[FX,FY] ⊠ FZ
τ
[FX,FX ⊠ [FX,FY] ⊠ FZ]
[1,π ⊠ 1]
[FX,FY ⊠ FZ]

Theorem 17: Let $F: \mathcal{C} \longrightarrow \mathcal{D}$ be a weakly monoidal functor. Assume that $\zeta: J \longrightarrow FI$ is an isomorphism and that $\zeta: F[P,I] \otimes FP \longrightarrow F([P,I] \otimes P)$ is an isomorphism for all finite objects $P \in \mathcal{C}$. If P is finite in $\mathcal{C}$ and if $[FP,-]$ exists in $\mathcal{D}$ then FP is finite in $\mathcal{D}$.

Proof: Since finiteness is equivalent to the fact that $i: I \longrightarrow [P,P]$ can be factored through $[P,I] \otimes P$ the following commutative diagram shows that $j: J \longrightarrow [FP,FP]$ can be factored through $[FP,J] \otimes FP$

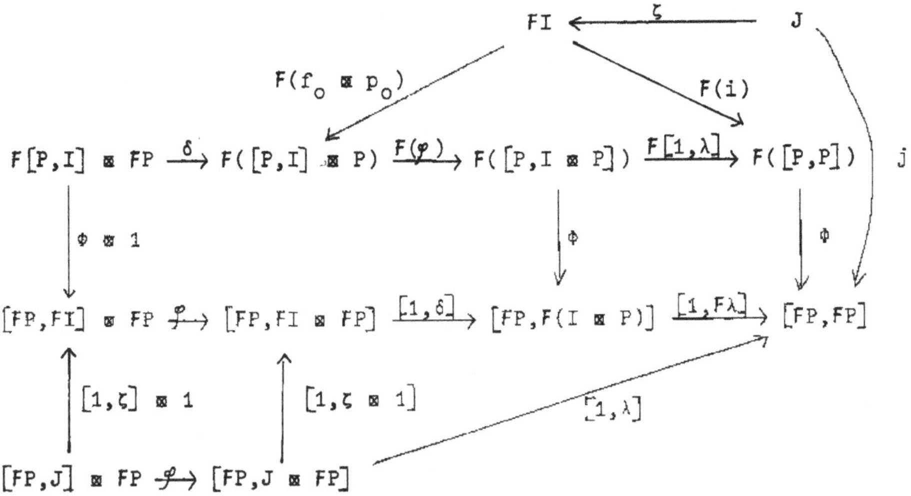

thus FP is finite.

Corollary 18: Under the assumption of Theorem 17 is the morphism $\Phi: F[P,X] \longrightarrow [FP,FX]$ an isomorphism for all $X \in \mathcal{C}$ and all finite $P \in \mathcal{C}$.

Proof: Let $f_o \otimes p_o: I \longrightarrow [P,I] \otimes P$ be the dual basis for P and $\bar{f}_o \otimes \bar{p}_o: J \longrightarrow [FP,J] \otimes FP$ be the dual basis for FP. Define $\Psi: [FP,FX] \longrightarrow F[P,X]$ to be

$$[FP,FX] \cong FI \otimes [FP,FX] \xrightarrow{F(f_o \otimes p_o) \otimes 1} F([P,I] \otimes P) \otimes [FP,FX] \xrightarrow{\delta^{-1} \otimes 1}$$
$$F[P,I] \otimes FP \otimes [FP,FX] \xrightarrow{1 \otimes \pi} F[P,I] \otimes FX \longrightarrow F[P,X] .$$

Omitting some of the obvious isomorphisms we get a commutative diagram

$$\begin{array}{ccccccc}
[FP,FX] & \xrightarrow{\delta^{-1} F(f_o \otimes p_o) \otimes 1} & F[P,I] \otimes FP \otimes [FP,FX] & \xrightarrow{1 \otimes \pi} & F[P,I] \otimes FX & \longrightarrow & F[P,X] \\
& \searrow{\scriptstyle \bar{f}_o \otimes \bar{p}_o \otimes 1} & \downarrow{\scriptstyle \Phi \otimes 1 \otimes 1} & & \downarrow{\scriptstyle \Phi \otimes 1} & & \downarrow{\scriptstyle \Phi} \\
& & [FP,FI] \otimes FP \otimes [FP,FX] & \xrightarrow{1 \otimes \pi} & [FP,FI] \otimes FX & \cong & [FP,FX]
\end{array}$$

where the left triangle commutes by the construction of $\bar{f}_o \otimes \bar{p}_o$ in Theorem 17 and right square commutes in the same way as the middle of the diagram in the proof of Theorem 17 does. If we look at the lower part of our diagram we see that the morphism $[FP,FX] \longrightarrow [FP,FX]$ is the identity since $\bar{f}_o\langle\bar{p}_o\rangle g = g$ for all $g \in [FP,FX](Y)$. The upper part is $\Phi\Psi$, hence $\Phi\Psi = \mathrm{id}$. Conversely the commutative diagram

$$\begin{array}{ccccccc}
F[P,X] & \longrightarrow & F[P,I] \otimes FP \otimes F[P,X] & \xrightarrow{1 \otimes \delta} & F[P,I] \otimes F(P \otimes [P,X]) & & \\
\downarrow{\scriptstyle \Phi} & & \downarrow{\scriptstyle 1 \otimes 1 \otimes \Phi} & & \downarrow{\scriptstyle 1 \otimes F(\pi)} & & \\
[FP,FX] & \longrightarrow & F[P,I] \otimes FP \otimes [FP,FX] & \xrightarrow{1 \otimes \pi} & F[P,I] \otimes FX & \longrightarrow & F[P,X]
\end{array}$$

shows $\Psi\Phi = \mathrm{id}$.

Corollary 19: *Under the assumptions of Theorem 17 if P is a progenerator then FP is a progenerator. If F preserves difference cokernels and P is faithfully projective then FP is faithfully projective.*

Proof: Let P be finite. P is a progenerator iff there is a morphism $f: I \longrightarrow P \otimes [P,I]$ such that

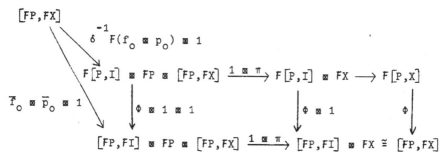

$$\begin{array}{ccc} & & I \\ & \swarrow^{f} & \downarrow 1 \\ P \boxtimes [P,I] & \xrightarrow{\pi} & I \end{array}$$

commutes. Now the diagram

$$\begin{array}{ccc} & & J \\ & \swarrow^{F(f)\zeta} & \wr\Vert\ \zeta \\ F(P \boxtimes [P,I]) & \xrightarrow{F(\pi)} & FI \\ \wr\Vert\ (1 \boxtimes [1,\zeta^{-1}])(1 \boxtimes \Phi)\delta^{-1} & & \wr\Vert\ \zeta^{-1} \\ FP \boxtimes [FP,J] & \xrightarrow{\pi} & J \end{array}$$

commutes hence FP is a progenerator. In the case of a faithfully projective P we have to replace

$P \boxtimes [P,I]$ by $P \underset{[P,P]}{\boxtimes} [P,I]$ and $FP \boxtimes [FP,J]$ by

$FP \underset{[FP,FP]}{\boxtimes} [FP,J] \cong F(P \underset{[P,P]}{\boxtimes} [P,I])$. The last isomorphism is a consequence of the fact that F preserves difference cokernels.

<u>Theorem</u> 20: <u>Let</u> $F: C \longrightarrow \mathcal{D}$ <u>be a weakly monoidal functor such that</u> $\zeta: J \longrightarrow FI$ <u>is an isomorphism</u>, $\delta: FX \boxtimes FP \longrightarrow F(X \boxtimes P)$ <u>is an isomorphism for all</u> $X \in C$ <u>and for all finite</u> $P \in C$, $[FP,-]$ <u>exists for all finite</u> $P \in C$ <u>and</u> F <u>preserves difference cokernels</u>. <u>Then</u> F <u>induces homomorphisms of Brauer groups</u> $B_i(F): B_i(C) \longrightarrow B_i(\mathcal{D})$ <u>for</u> $i = 1, 2$ <u>such that</u>

$$\begin{array}{ccc} B_2(C) & \xrightarrow{B_2(F)} & B_2(\mathcal{D}) \\ \downarrow \xi & & \downarrow \xi \\ B_1(C) & \xrightarrow{B_1(F)} & B_1(\mathcal{D}) \end{array}$$

<u>commutes</u>.

Proof: Let A be a monoid in C . Then FA is a monoid in $\mathcal{D}$ with the multiplication $FA \boxtimes FA \xrightarrow{\delta} F(A \boxtimes A) \xrightarrow{F(\mu)} FA$ and unit $J \xrightarrow{\zeta} FI \xrightarrow{F(\eta)} FA$. If A is i-Azumaya, $i = 1, 2$, then FA is faithfully projective resp. a progenerator by Corollary 19, Proposition 1 and Theorem 12. So we only have to show that

$\psi: FA \boxtimes FA \longrightarrow [FA,FA]$ is an isomorphism. ψ is induced by

$T: A \boxtimes A \boxtimes A(X) \ni a \boxtimes b \boxtimes c \longmapsto bac \in A(X)$, so that $\psi = \chi(T)$ where $\chi: C(X \boxtimes Y,Z) \cong C(Y,[X,Z])$.

Now the diagram

$$\begin{array}{ccc} FA \boxtimes FA & \xrightarrow{\chi(T)} & [FA,FA] \\ \downarrow \delta & & \uparrow \Phi \\ F(A \boxtimes A) & \xrightarrow{F(\chi(T))} & F[A,A] \end{array}$$

commutes, since

$$\begin{array}{ccc} FA \boxtimes (FA \boxtimes FA) & \xrightarrow{T} & FA \\ \downarrow 1 \boxtimes \delta & & \uparrow F(T) \\ FA \boxtimes F(A \boxtimes A) & \xrightarrow{\delta} & F(A \boxtimes (A \boxtimes A)) \end{array}$$

commutes so that by applying χ we get

$\Phi \cdot F(\chi(T))\cdot\delta = \chi(F(T)\cdot\delta)\cdot\delta = [1,F(T)]\circ\chi(\delta)\circ\delta = \chi(T)$. The first identity results from the commutativity of

$$\begin{array}{ccccc} C(X \boxtimes Y,Z) & \xrightarrow{F} & \mathcal{D}(F(X \boxtimes Y)\ ,\ FZ) & \xrightarrow{\mathcal{D}(\delta,1)} & \mathcal{D}(FX \boxtimes FY,FZ) \\ \downarrow \chi & & & & \downarrow \chi \\ C(Y,[X,Z]) & \xrightarrow{F} & \mathcal{D}(FY,F[X,Z]) & \xrightarrow{\mathcal{D}(1,\Phi)} & \mathcal{D}(FY,[FX,FZ]) \end{array} .$$

Now $\Phi: F[A,A] \longrightarrow [FA,FA]$ is an isomorphism by Corollary 18 and $F(\chi(T)) = F(\psi)$ is an isomorphism since A is Azumaya. Since δ is an isomorphism, too, we get that $\psi = \chi(T): FA \boxtimes FA \longrightarrow [FA,FA]$ is an isomorphism.

If $P \in C$ is faithfully projective or a progenerator in C then as above FP is faithfully projective or a progenerator in $\mathcal{D}$ and $F[P,P] \cong [FP,FP]$ as monoids using the first commutative diagram we proved for Φ .

Thus if A and B are i-Azumaya, then FA and FB are i-Azumaya. If A and B are equivalent w.r.t. $B_i(C)$, then so are FA and FB . Finally we have $F(A \boxtimes B) \cong FA \boxtimes FB$ and $FI = J$ so that F induces homomorphisms $B_i(F): B_i(C) \longrightarrow B_i(\mathcal{D})$ such that the diagram in the theorem commutes.

If $F: C \longrightarrow \mathcal{D}$ is a functor satisfying the conditions of Theorem 20 then we define the kernel of $B_i(F)$ as $B_i(C,F)$ so that we get exact sequences

$$0 \longrightarrow B_i(C,F) \longrightarrow B_i(C) \longrightarrow B_i(\mathcal{D})$$

for $i = 1, 2$. $B_i(C,F)$ containes those elements $[A]$ of $B_i(C)$ with $[FA] = [[P,P]]$ for some $P \in \mathcal{D}$ which is faithfully projective resp. a progenerator. These i-Azumaya monoids A are called F-split. From Theorem 20 follows immediatly a homomorphism

$$\xi\colon B_2(C,F) \longrightarrow B_1(C,F) .$$

If C is a symmetric monoidal closed category with difference kernels and difference kokernels and $K \in C$ is commutative monoid, then ${}_KC$ is again a symmetric monoidal closed category with $\otimes_K$ as tensor product and K as basic object. Then the functor $C \ni X \longmapsto K \otimes X \in {}_KC$ has all properties required in Theorem 20 hence there are homomorphisms $B_i(C) \longrightarrow B_i({}_KC)$ with kernels $B_i(K/C)$.

References:

[1] Auslander, M. and Goldmann, O.: The Brauer group of a commutative ring, Trans.Amer.Math.Soc. 97 (1960), 367 - 409.

[2] Long, F.W.: The Brauer group of dimodule algebras, J. of Algebra 30 (1974), 559 - 601.

[3] MacLane, S.: Categories for the working mathematician, Graduate Texts in Mathematics. Springer New York - Heidelberg - Berlin 1971.

[4] Orzech, M. and Small, Ch.: The Brauer group of commutative rings. Leisure notes in Pure and Applied Mathematics. Marcel Dekker New York 1975.

[5] Pareigis, B.: Non-additive ring and module theory I: General theory of monoids.
To appear in: Publicationes Mathematicae Debrecen.

[6] Pareigis, B.: Non-additive ring and module theory II: C-catego - ries, C-functors and C-morphisms.
To appear in: Publicationes Mathematicae Debrecen.

[7] Pareigis, B.: Non-additive ring and module theory III: Morita theorems over monoidal categories.
To appear in: Publicationes Mathematicae Debrecen.

[8] Pareigis, B.: Non-additive ring and module theory V: Projective and flat objects.
To appear in: Algebra-Berichte.

[9] Fisher-Palmquist, J.: The Brauer group of a closed category, Proc.Amer.Math.Soc. 50 (1975), 61 - 67.

Brauer Groups of Graded Algebras

Morris Orzech

My talk at the Conference presented a survey of work on Brauer groups of algebras with gradings and actions. I shall not resurrect this approach here, although I shall maintain contact with its spirit. The aim of presenting an overview will be adhered to in the course of doing other things, viz. presenting some new results relating to Long's Brauer groups of module and dimodule algebras (§1), proposing two generalizations of these groups (§2), and suggesting questions raised by the work to date in the area (§3). We refer the reader to the diagram preceding the references as a possible aid in recalling the context of our discussion and digressions.

§1. Some computations. Let R be a commutative ring. Recall that the Brauer groups of module algebras (or comodule algebras, or dimodule algebras) defined by Long in [8] are constructed relative to a commutative cocommutative finite (i.e. finitely generated projective) Hopf R-algebra H. An H-dimodule (resp. an H-dimodule algebra) M is an R-module (resp. an R-algebra) together with action and coaction maps

$$\alpha : M \to M \otimes H \ , \ \gamma : H \otimes M \to M$$

which are R-module (resp. R-algebra) maps satisfying axioms of associativity, coassociativity, unitarity and counitarity. If α(resp. γ) is trivial then M is an H-module (resp. H-comodule). For M an H-dimodule which is a faithfully projective R-module $\mathrm{End}_R(M)$ is an H-dimodule algebra. The category of H-dimodule algebras is closed under smash product $\#$, and the Brauer group BD(R,H) of H-dimodule algebras is defined using $\#$. The Brauer group of H-module (resp. H-comodule) algebras is denoted by BM(R,H) (resp. BC(R,H)). We begin by stating a recent result of Beattie [2]:

Theorem 1. Let R be a commutative ring, H a commutative, cocommutative Hopf R-algebra which is a finitely generated projective R-module. There is then a split exact sequence

$$0 \to B(R) \to BM(R,H) \to Gal(R,H) \to 1$$

where Gal(R,H) is the group of Galois H-objects defined in [3].

This complements the main result of [10], which obtains a sequence such as the above for BC(R,RG) (G a finite abelian group) replacing BM(R,H). The proof in [2] is analogous to that in [10] but is naturally complicated by passage from RG to a Hopf algebra H and necessarily to Galois H-objects.

The next theorem extends results of [7] and [9]:

Theorem 2. Let n be a square-free integer, C_n a cyclic group of order n. Let $R \subseteq S$ be an inclusion of commutative rings satisfying:

(1) R contains $1/n$ and a primitive n-th root of 1;

(2) The n-torsion part of $Pic(R)$ is trivial;

(3) S is a faithfully flat R-module;

(4) $B(S)$, the Brauer group of S, is trivial and $U(S)/U(S)^n$ is trivial.

Then there are exact sequences:

$$1 \to N \to BD(R,C_n) \to BD(S,C_n) \to 1$$

$$1 \to B(R) \to N \to R_n \times R_n \to 1$$

where $R_n = U(R)/U(R)^n$ and the map from $BD(R,C_n)$ to $BD(S,C_n)$ sends (A) to $(S \otimes_R A)$.

Proof. The statement of this theorem is taken almost verbatim from Theorem 3.1 of [7]. The proof there works with the hypotheses we have put on R and S provided the results of §5 of [9] replace those from §2 of [7]. The argument used in Corollary 3.12 of [7] (or the above sequences applied directly) can be used to obtain:

multiplication via

$$\begin{array}{ccccccc} \bar{A}\otimes\bar{A} & \xrightarrow{\alpha\otimes 1} & A\otimes H\otimes A & \xrightarrow{1\otimes\Psi\otimes 1} & A\otimes J\otimes A & & \sigma = \text{switch}, \\ \downarrow & & & & \downarrow & & \\ \bar{A} & \xleftarrow[m]{} & A\otimes A & \xleftarrow[\sigma]{} & A\otimes A & & m = \text{multiplication}. \end{array}$$

It is just as straightforward to modify the usual multiplication on $A\#B$ by using Ψ, to define $A\#B$ in our context. We obtain a group we'll call $Br(\Psi)$. We can recover Long's groups easily : let $1_H : H \to H$, $i_H : R \to H$, $\varepsilon_H : H \to R$. Then

$$B(1_H) = BD(R,H);\ B(i_H) = BM(R,H);\ B(\varepsilon_H) = BC(R,H).$$

For G a finite abelian group and $\phi : G \times G \to U(R)$ a bilinear map we can recover $B_\phi(R,G)$ as a direct summand of a $Br(\Psi)$. Define $\phi^* : RG \to GR$ by $\phi^*(\sigma)(\tau) = \phi(\sigma,\tau)$. We then have a map from $B_\phi(R,G)$ to $Br(\phi^*)$ sending a G-graded R-algebra A, (i.e. an RG-comodule relative to $\alpha : A \to A \otimes RG$) to

$$\begin{array}{ccc} GR\otimes A & \xrightarrow{\gamma} A \xrightarrow{\alpha} & A\otimes RG \\ {\scriptstyle 1\otimes\alpha}\downarrow & \Vert & \\ GR\otimes A\otimes RG & \xrightarrow[ev]{} A & \end{array} \qquad ev(h\otimes a\otimes\sigma) = h(\sigma)a.$$

This map is split by the "forgetful functor" which drops the GR-action on a ϕ^*-algebra, so $B_\phi(R,G)$ is a direct summand of $B(\phi^*)$.

We shall now consider functorial properties of $Br(\Psi)$. Consider a commutative diagram of Hopf algebras and Hopf algebra maps, which we shall call "a map $f : \Psi \to \Psi'$ " :

$$\begin{array}{ccc} H & \xrightarrow{\Psi} & J \\ {\scriptstyle f_*}\downarrow & & \downarrow{\scriptstyle f^*} \\ H' & \xrightarrow[\Psi']{} & J' \end{array}$$

We get a functor from Ψ-modules to Ψ'-modules in an obvious way, and a homomorphism $Br(f) : Br(\Psi) \to Br(\Psi')$. Using the fact that $Br(f)$ can be shown to be monic, it is possible to recover two observations made by Long in [7] .

<u>Proposition 1</u>. The collection of commutative diagrams implicit in the following picture

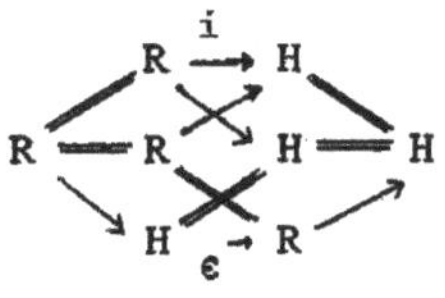

gives rise to inclusions

$$\begin{array}{ccccc} & & BM(R,H) & & \\ & \subseteq & & \subseteq & \\ B(R) & & & & BD(R,H) \\ & \subseteq & & \subseteq & \\ & & BC(R,H) & & \end{array}$$

Proposition 2. Let $\Psi_i : H_i \to J_i$ be a map of Hopf algebras, $i = 1,2$. Then the commutative diagram

$$\begin{array}{ccc} H_1 & \xrightarrow{\Psi_1} & J_1 \\ {\scriptstyle 1\otimes i}\downarrow & & \downarrow{\scriptstyle 1\otimes\varepsilon} \\ H_1 \otimes H_2 & \xrightarrow[\Psi]{} & J_1 \otimes J_2 \end{array} \qquad \Psi = \Psi_1 \otimes \Psi_2$$

induces an inclusion $Br(\Psi_1) \to Br(\Psi)$.

Long proves this as Theorem 1.8 of [7] for $J_i = RG_i = H_i$, G_i a finite abelian group.

§3 Some questions. Our first two questions refer to §2.

(1) Let H be a Hopf algebra as in §2 and $J = H \otimes H^*$. Let $\theta : J \to J^*$ be the identity map. A θ-module M has an H-dimodule structure arising from the action by J and an H-dimodule structure arising from the coaction by J. Axiom (*) says these two H-dimodule structures agree. It is not clear that $BD(\theta) = BD(R,H)$ because the smash product of θ-algebras defined via J does not coincide with their smash product as H-dimodule algebras. Is $BD(\theta) = BD(R,H)$?

(2) We showed in §2 that $B_{\phi}(R,G)$ is a direct summand of $Br(\phi^*)$. Is it all of $Br(\phi^*)$?

(3) in [1], [5], [11] and [12] the treatment of graded Azumaya algebras is via a Morita theory which permits two equivalent characterizations of A being "Azumaya", viz. A is "central" and "separable" (in a suitable sense), or $A \# \bar{A} \to \mathrm{End}(A)$ is an isomorphism, etc. Only the second of these characterizations plays a role in [7] and [8]. Is a treatment of separability, analogous to that of [5], possible in the setting of [8]?

(4) The work in [4] suggests relating BD(R,H) to suitable Galois extensions. The same work suggests that such Galois extensions would be quite messy. Are either of these suggestions capable of being pursued?

In the following schematic representation G denotes a finite abelian group, Γ an abelian group, H a Hopf algebra, A the kind of R-algebra being considered in the work described, X the notation for the Brauer group defined. The pattern of connecting lines has a chronological and/or philosophical validity which has been tempered by a desire for visual symmetry.

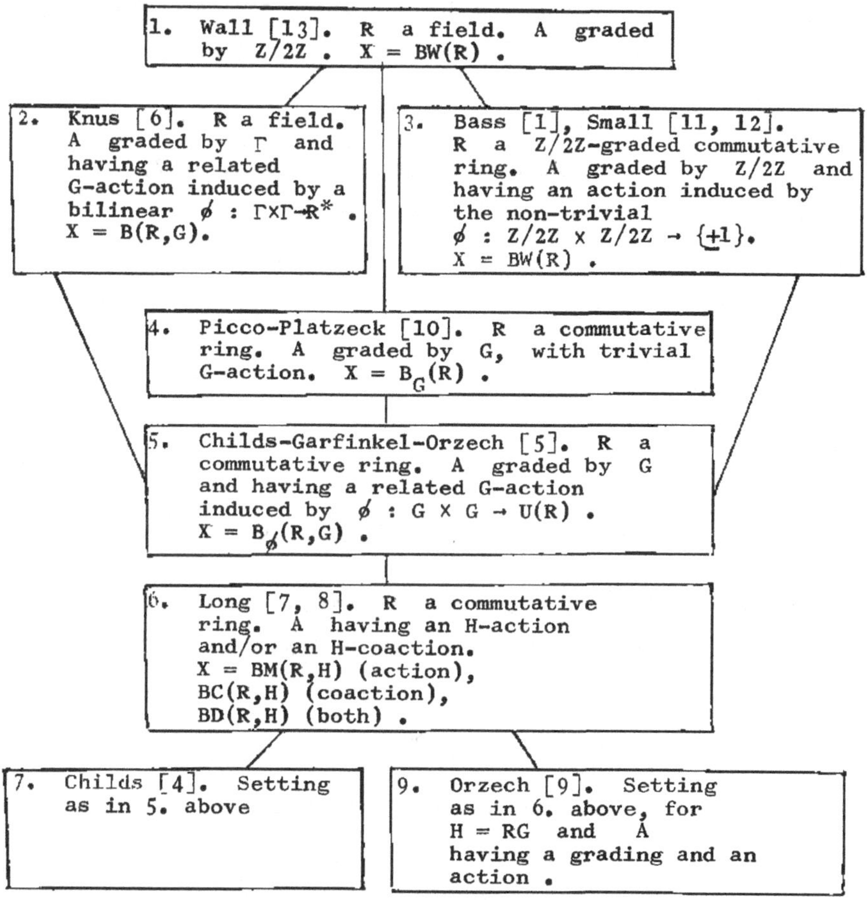
1. Wall [13]. R a field. A graded by Z/2Z . X = BW(R) .
2. Knus [6]. R a field. A graded by Γ and having a related G-action induced by a bilinear ϕ : Γ×Γ→R* . X = B(R,G).
3. Bass [1], Small [11, 12]. R a Z/2Z-graded commutative ring. A graded by Z/2Z and having an action induced by the non-trivial ϕ : Z/2Z x Z/2Z → {±1}. X = BW(R) .
4. Picco-Platzeck [10]. R a commutative ring. A graded by G, with trivial G-action. X = B_G(R) .
5. Childs-Garfinkel-Orzech [5]. R a commutative ring. A graded by G and having a related G-action induced by ϕ : G × G → U(R) . X = B_ϕ(R,G) .
6. Long [7, 8]. R a commutative ring. A having an H-action and/or an H-coaction. X = BM(R,H) (action), BC(R,H) (coaction), BD(R,H) (both) .
7. Childs [4]. Setting as in 5. above
9. Orzech [9]. Setting as in 6. above, for H = RG and A having a grading and an action .

Fig. 1. Overview of Brauer Groups of Structured Algebras

References

1. H. Bass, Lectures on Topics in Algebraic K-Theory, Tata Institute for Fundamental Research, Bombay, 1967.

2. M. Beattie, A direct sum decomposition for the Brauer group of H-module algebras, J. Algebra, to appear.

3. S.U. Chase, and M.E. Sweedler, Hopf algebras and Galois Theory, Lecture Notes in Mathematics 97, Springer-Verlag, Berlin, 1969.

4. L.N. Childs, The Brauer group of graded algebras II : graded Galois extensions, Trans. Amer. Math. Soc. 204 (1975), 137-160.

5. ———, G. Garfinkel and M. Orzech, The Brauer group of graded Azumaya algebras, Trans. Amer. Math. Soc. 175 (1973), 299-326

6. M.-A. Knus, Algebras graded by a group, Category Theory, Homology Theory and their Applications II, Lecture Notes in Mathematics 92, Springer-Verlag, Berlin, 1969 .

7. F.W. Long, A generalization of the Brauer group of graded algebras, Proc. London Math. Soc. (3) 29 (1974), 237-256 .

8. ———, The Brauer group of dimodule algebras, J. Algebra 30 (1974), 559-601 .

9. M. Orzech, On the Brauer group of modules having a grading and an action, Canad. J. Math., to appear.

10. D.J. Picco and M.I. Platzeck, Graded algebras and Galois extensions, Rev. Un. Mat. Argentina 25 (1971), 401-415 .

11. C. Small, The Brauer-Wall group of a commutative ring, Trans. Amer. Math. Soc. 156 (1971), 455-491.

12. ————, The group of quadratic extensions, J. Pure Applied Alg. 2 (1972), 83-105 .

13. C.T.C. Wall, Graded Brauer groups, J. Reine Angew. Math. 213 (1964), 187-199 .

ON A VARIANT OF THE WITT AND BRAUER GROUPS

by

Stephen U. Chase*

The purpose of this note is to develop further an analogy, elucidated in [7], between quadratic forms over a field k of characteristic not two and principal homogeneous spaces for affine k-group schemes. This analogy can be made precise by means of graded group schemes. Namely, the quadratic forms on a k-space V of finite dimension are in bijective correspondence with what are essentially principal homogeneous spaces for the graded k-group scheme represented by the exterior algebra of V. Under this bijection, the hyperbolic quadratic form on $V \oplus V^*$ [13, p. 122], where $V^* = \mathrm{Hom}_k(V,k)$, corresponds to a certain twisted or "smash" product of principal homogeneous spaces which generalizes the classical construction of cyclic algebras.

In this paper we exploit this analogy to construct, for principal homogeneous spaces arising from ordinary (i.e., ungraded) k-group schemes, a counterpart of the Witt group of non-degenerate quadratic forms over k as described, for example, in Lam [14, Ch. Two, §1, pp. 34-37]. Of course, it is well-known that the Witt group of quadratic forms carries a ring structure, but such a structure appears unlikely to

*This work was supported in part by NSF MPS73-04876.

exist in our context. We also define a homomorphism of our Witt group $Z(k)$ into the (ordinary, ungraded) Brauer group $Br(k)$ of k. This homomorphism is surjective for local and global fields, and our hope is that, for such fields, $Z(k)$ will yield a useful and interesting arithmetical invariant. So far, however, our results are (to say the least) fragmentary, and we state them here with only the barest indications of proof. Some of the details have appeared in [7], and others will appear elsewhere.

We now describe, more precisely and in greater detail, the concepts with which we shall work. If G is a k-group scheme acting on a k-scheme X, then X is called a "principal homogeneous space" (PHS) for G if G acts freely and transitively on X in a sense easily made precise (see, e.g., Waterhouse [23], Chase [6, §5], or the paragraph following (1.3) below). If $G = \text{Spec}\,(C)$ and $X = \text{Spec}\,(A)$ are affine, then the definition of a PHS can be phrased as an algebraic condition on A, and one then notes that this condition is meaningful even if A is not commutative (in which case the scheme X no longer exists). This leads us to the notion of a "pseudo-principal homogeneous space" (PPHS) for G, a concept which has appeared in the literature in various guises and special cases (see, e.g., Harrison [11], Hoechsmann [12], Chase-Rosenberg [4], and Orzech [17]).

Now let $G = \text{Spec}\,(C)$ be a finite commutative k-group scheme, and $G^D = \text{Spec}(H)$ be its Cartier dual [8, pp. 156-159], with $H = C^*$ the dual Hopf algebra of G. If G and G^D act on the affine k-schemes $X = \text{Spec}(A)$ and $Y = \text{Spec}(B)$, respectively, Gamst-Hoechsmann [10] have defined a k-algebra A#B, called the smash product of A and B, which contains A and B as subalgebras and which generalizes the classical construction of quaternion and cyclic algebras [1, pp. 82-83]. $A\#B = A\otimes_k B$ as k-spaces, with a "twisted" multiplication. If X and Y are PHS's for G and G^D, respectively, then A#B is a central simple k-algebra, and the smash product construction then gives rise to a pairing $\underline{\overline{X}}(G)\otimes\underline{\overline{X}}(G^D) \to Br(k)$, with Br(k) the Brauer group of k and $\underline{\overline{X}}(G)$ the abelian group of isomorphism classes of PHS's for G, etc. [23, p. 181]. As is proved in [10], this pairing coincides with a cohomological cup product pairing treated by Schatz [18], special cases of which are well-known in class field theory and yield the norm residue symbol for local fields. Actually, in order to define A#B, one needs only k-algebras A and B which are acted upon by G and G^D, respectively; the requirement that A and B be commutative (and hence correspond to affine k-schemes) is unnecessary.

In this paper, the introductory §1 presents certain basic definitions and elementary facts regarding affine group schemes and their actions on schemes and algebras. In §2 we introduce and discuss PPHS's and their smash products. The main theorem of the section asserts that, if G is a finite commutative k-group scheme and A, B are PPHS's for G and G^D, respectively, then A#B is, in a natural way, a PPHS for $G\times G^D$. This theorem provides many examples of (non-commutative) PPHS's, since A#B is usually not commutative even if A and B are.

In §3 we exhibit a bijective correspondence between isomorphism classes of PPHS's for a finite commutative k-group scheme G and isomorphism classes of central presheaf extensions of G^D by the multiplicative k-group scheme G_m. A similar bijection relating PHS's for G and abelian presheaf extensions is explicated in Chase-Sweedler [5, Ch. III, pp. 84-126], Schatz [20], and Waterhouse [23], and provides a group-scheme-theoretic generalization of the classical Kummer theory for fields.

The relation between PPHS's and central extensions just mentioned has the following pleasant feature. Let G be a finite commutative k-group scheme, and A,B be PPHS's for G and G^D, respectively. Let ξ_A be the central extension of G^D by G_m corresponding to A, and $\overline{\xi}_A$ be the central extension of $G^D \times G$ by G_m obtained from ξ_A by pull-back along the projection $G^D \times G \to G^D$; define similarly ξ_B and $\overline{\xi}_B$. We observe in §3 that the isomorphism class of the central extension of $(G \times G^D)^D = G^D \times G$ by G_m corresponding to the PPHS $A\#B$ is

$$cl(\overline{\xi}_A) + cl(\overline{\xi}_B) + cl(\Xi)$$

where "cl" means "isomorphism class of", "+" denotes the addition in the abelian group of central extensions of $G^D \times G$ by G_m, and-

$$\Xi\colon 1 \to G_m \to E \to G^D \times G \to 1$$

is a central extension which can be roughly described by the statement that $E = G_m \times G^D \times G$ as a k-scheme, but the group multiplication is "twisted" by means of the Cartier duality pairing $G^D \times G \to G_m$ [8, pp. 156-159] in a manner which generalizes the construction of the Heisenberg group. These extensions, which we call "Heisenberg extensions", appear in Weil [24, Ch. I, §4, p. 149] in the context of locally compact abelian groups.

In §4, pursuing the afore-mentioned analogy which motivates this work, we introduce counterparts for PPHS's of non-degenerate and metabolic quadratic forms [13, p. 122], and define our variant $Z(k)$ of the Witt group as, essentially, a Grothendieck group of non-degenerate PPHS's for finite commutative k-group schemes, modulo the subgroup generated by the metabolic PPHS's. We also construct and briefly discuss the homomorphism $\alpha_k\colon Z(k) \to Br(k)$. In §5 we define a second homomorphism $\omega_k\colon Z(k) \to W_{sp}(k)$, where $W_{sp}(k)$ is an analogue of the Witt group constructed from pairings of finite commutative k-group schemes into G_m which are symplectic in an obvious sense. The existence of this mapping arises from the correspondence between PPHS's and central extensions of group schemes presented in §3.

Throughout this paper we shall make use of the following conventions and notation. All rings will have units, and a subring will always be assumed to have the same unit as the larger ring. If R is a ring we denote by $U(R)$ the multiplicative group of invertible elements of R. If V is a vector space over a field k, $[V:k]$ denotes the k-dimension of V. We shall use without reference the terminology and basic notions of category theory.

1. Affine Group Schemes and Principal Homogeneous Spaces

We shall briefly discuss some elementary facts regarding affine group schemes, their actions on schemes and algebras, and their corresponding principal homogeneous spaces. For a detailed treatment see, e.g., Demazure-Gabriel [8, Chapitres I-III], or Sweedler [21] for a Hopf algebraic approach. Some of this material is outlined also in Chase [6, §'s 1, 5, and 6], and we shall find it convenient to adopt the terminology and notation of that paper. We deal only with affine schemes over a fixed base field k, and shall write $\times = \times_{\mathrm{Spec}(k)}$ (for k-schemes) and $\otimes = \otimes_k$ (for k-spaces). If $X = \mathrm{Spec}(A)$ is a k-scheme and T is a commutative k-algebra, we denote by $X(T) = \mathrm{Alg}_k(A,T)$ the set of T-valued points of X; this is simply the set of k-scheme morphisms $\mathrm{Spec}(T) \to X$. If $G = \mathrm{Spec}(C)$ is a k-group scheme, then C is a commutative Hopf k-algebra (i.e., involutive bialgebra), and G(T) is a group for each T as above; the resulting group-valued functor will also be denoted by G.

A k-group scheme $G = \mathrm{Spec}(C)$ acts on a k-scheme X if there is given a k-scheme morphism

$$(1.1) \qquad \varphi : X \times G \to X$$

which is associative and unitary in the obvious sense. If $X = \mathrm{Spec}(A)$, we shall often say simply that G acts on A. In that case, $\varphi = \mathrm{Spec}(\theta)$, where $\theta : A \to A \otimes C$ is a k-algebra homomorphism which gives to A the structure of a C-comodule. Moreover, if T is a commutative k-algebra, then G(T) operates on $A \otimes T$ by T-algebra automorphisms according to the formula -

$$(1.2) \qquad \sigma(a\otimes t) = \{1_A \otimes \sigma')(\theta(a))\}(1\otimes t)$$

for a in A and t in T, where $\sigma' : C \to T$ in $\mathrm{Alg}_k(C,T)$ corresponds to the element σ of $G(T)$. This operation is natural in T.

If, conversely, $G(T)$ operates on $A\otimes T$ by T-algebra automorphisms in a manner functorial in T, then we obtain an action of G on $X = \mathrm{Spec}(A)$ as follows. Let ω be the element of $G(C) = \mathrm{Alg}_k(C,C)$ corresponding to the identity map 1_C of C, and define $\theta : A \to A\otimes C$ by the formula $\theta(a) = \omega(a\otimes 1)$ for a in A. Then θ is a k-algebra homomorphism, and applying the functor Spec yields the desired k-scheme morphism $\varphi : X\times G \to X$.

If G acts on X as above, then $\varphi : X\times G\to X$, together with the projection $\pi : X\times G \to X$, yield a k-scheme morphism $(\pi,\varphi) : X\times G \to X\times X$. If T is a commutative k-algebra, then the induced map on T-valued points sends (x,σ) in $X(T)\times G(T)$ to $(x,x\sigma)$ in $X(T)\times X(T)$. $(\pi,\varphi) = \mathrm{Spec}(\gamma)$, where $\gamma : A\otimes A \to A\otimes C$ satisfies the formula -

$$(1.3) \qquad \gamma(a\otimes a') = (a\otimes 1)\theta(a')$$

for a,a' in A [6, (1.10b)]. Note that (π,φ) is an A-scheme morphism if $X\times X$ is viewed as a scheme over X (i.e., an A-scheme) via projection on the left factor, and so γ is, in a similar fashion, a homomorphism of A-algebras. X will be called a principal homogeneous space (PHS) for G if (π,φ) is an isomorphism of schemes; i.e. γ is bijective (see, e.g., [6, §5];

other references are [23], or [5, Ch. I, §4, pp.28-40], where a PHS is called a "Galois object"). Note that G is a PHS for itself via "right multiplication".

Example 1.4. Let Γ be a finite (abstract) group. The constant k-group scheme corresponding to Γ is the k-group scheme $G = \mathrm{Spec}(k^\Gamma)$, where k^Γ is the k-algebra of all functions from Γ to k, the algebra operations being defined pointwise and the k-coalgebra structure maps -

$$k^\Gamma \to k^{\Gamma \times \Gamma} = k^\Gamma \otimes k^\Gamma$$

$$k^\Gamma \to k$$

and antipode [21, p.71] -

$$k^\Gamma \to k^\Gamma$$

being induced by the multiplication in Γ, evaluation at 1, and induced by the antipodal map $\sigma \to \sigma^{-1}$, respectively. If T is a commutative k-algebra, then the mapping $\Gamma \to \mathrm{Alg}_k(k^\Gamma, T) = G(T)$, where σ in Γ goes to "evaluation at σ", is a homomorphism of groups, is natural in T, and is an isomorphism if T is connected. An action of G on $X = \mathrm{Spec}(A)$ is equivalent to an operation of Γ on A by k-algebra automorphisms, the two concepts being related by the formula -

(1.5) $\theta(a)(\sigma) = \sigma(a)$ (a in A, σ in Γ)

with $\theta : A \to A \otimes k^{\Gamma} = A^{\Gamma}$ as in (1.2) and the preceding discussion. If A is a field, then it is easy to see that X is a PHS for G if and only if A is a normal separable extension of k with Galois group Γ [5, Ch. II, p.59].

Example 1.6. If Γ is an (abstract) group, not necessarily finite, then $k\Gamma$, the group algebra of Γ with coefficients in k, is a cocommutative Hopf k-algebra with coalgebra structure maps -

$$k\Gamma \to k\Gamma \otimes k\Gamma$$

$$k\Gamma \to k$$

the usual diagonal and augmentation maps (i.e., $\sigma \to \sigma \otimes \sigma$ and $\sigma \to 1$, respectively, for all σ in Γ), and antipode

$$k\Gamma \to k\Gamma$$

where $\sigma \to \sigma^{-1}$. If, in addition, Γ is abelian, then $G = \mathrm{Spec}(k\Gamma)$ is a commutative k-group scheme. If G acts on $X = \mathrm{Spec}(A)$ and σ is in Γ, we define a subset $A_\sigma \subseteq A$ by the condition -

$$A_\sigma = \{a \text{ in } A \mid \theta(a) = a \otimes \sigma\}$$

with $\theta : A \to A \otimes k\Gamma$ arising as above from the action of G on X. It is then easy to see that each A_σ is a k-subspace of A, $A = \coprod_\sigma A_\sigma$, and $A_\sigma A_\tau \subseteq A_{\sigma\tau}$; i.e., A is a Γ-graded k-algebra. Conversely, if A is a Γ-graded k-algebra, then we obtain an action of G on $X = \mathrm{Spec}(A)$ if we define $\theta : A \to A \otimes k\Gamma$ by the formula $\theta(a) = a \otimes \sigma$ for a in A_σ.

In particular, suppose that Γ is a cyclic group of order n with generator σ. If T is a commutative k-algebra, then evaluation at σ yields a group homomorphism $G(T) = \mathrm{Alg}_k(k\Gamma, T) \to U(T)$ which maps $G(T)$ isomorphically onto the subgroup $\mu_n(T) = \{\zeta \text{ in } T \mid \zeta^n = 1\}$ of "n^{th} roots of 1 in T", and we identify $G(T)$ with $\mu_n(T)$ via this natural isomorphism (one then usually writes $G = \mu_n$). If A is as above, then the operation of $\mu_n(T)$ on $A \otimes T$ by T-algebra automorphisms satisfies the formula $\zeta(a \otimes t) = a \otimes \zeta^i t$ for ζ in $\mu_n(T)$, t in T, and a in A_{σ^i}. For further remarks on this case and a description of the resulting PHS's, see [5, Ch. I, pp.36-40].

2. Pseudo-Principal Homogeneous Spaces

The notion of a pseudo-principal homogeneous space arises from the observation (which is not new) that the definition of a PHS makes perfect sense without the requirement that A be a commutative k-algebra. Namely, if $G = \mathrm{Spec}(C)$ is a k-group scheme and A is a (not necessarily commutative) k-algebra, an action of G on A is defined to be an operation, for each commutative k-algebra T, of the group $G(T)$ on $A \otimes T$ by T-algebra automorphisms, this operation being natural in T. Such an action corresponds, as in (1.2), to a k-algebra homomorphism $\theta: A \to A \otimes C$ wich renders A a C-comodule. A will be called a pseudo-principal homogeneous space (PPHS) for G if the mapping $\gamma: A \otimes A \to A \otimes C$ of (1.3) (no longer a homomorphism of rings) is bijective. For further information on these objects, at least for the special case in which G is a constant group scheme, see [12], [4], or [17], where a PPHS is called a "non-commutative Galois extension". It is easy to see that the remarks regarding PHS's in (1.4) and (1.6) also hold for PPHS's.

Example 2.1. Let A_i be PPHS's for the k-group schemes $G_i = \mathrm{Spec}(C_i)$, $i = 1,2$, and set $G = G \times G_2 = \mathrm{Spec}(C_1 \otimes C_2)$ and $A = A_1 \otimes A_2$. Then the k-group scheme G acts on the k-algebra A if, for each commutative k-algebra T, we require that $G(T)$ operate on $A \otimes T$ by T-algebra automorphisms according to the formula

$$(\sigma_1, \sigma_2)(a_1 \otimes a_2 \otimes t) =$$

$$\{(\eta_1 \otimes 1_T)(\sigma_1(a_1 \otimes 1))\}\{(\eta_2 \otimes 1_T)(\sigma_2(a_2 \otimes 1))\}(1 \otimes 1 \otimes t)$$

for a_i in A_i, σ_i in $G_i(T)$ (whence (σ_1,σ_2) is in $G(T)$) and $\eta_i : A_i \to A$ the obvious k-algebra homomorphisms. It is easy to see, by explicitly computing the homomorphisms $\theta : A \to A\otimes(C_1\otimes C_2)$ and $\gamma : A\otimes A \to A\otimes(C_1\otimes C_2)$, that the latter is an isomorphism and hence A is a PPHS for G.

Example 2.2. Assume that the characteristic of k is not two, let $a,b \neq 0$ be elements of k, and let A be the corresponding "quaternion algebra" over k; i.e., A is generated as a k-algebra by the elements α and β subject only to the relations $\alpha^2 = a$, $\beta^2 = b$, $\beta\alpha = -\alpha\beta$. Let Γ be the group of two elements; then $\Gamma\times\Gamma$ operates on A by k-algebra automorphisms according to the formulae -

$$\rho(\alpha) = \mathrm{sgn}(\sigma)\alpha \qquad \rho(\beta) = \mathrm{sgn}(\tau)\beta$$

for $\rho = (\sigma,\tau)$ in $\Gamma\times\Gamma$, with σ, τ in Γ and $\mathrm{sgn}:\Gamma \overset{\approx}{\to} \{+1,-1\} \subseteq \dot{k} = k - \{0\}$ an isomorphism of groups. By (1.6), this operation corresponds to an action on A of the constant k-group scheme G corresponding to $\Gamma\times\Gamma$. In the following discussion we shall state a general result which implies that A is a PPHS for G. First, however, we shall need some preliminary notions.

A k-group scheme $G = \mathrm{Spec}(C)$ will be called finite if C is of finite k-dimension, and in this paper we shall be primarily concerned with such group schemes. If G is finite, we write $[G:k] = [C:k]$, the rank of G over k. In that case $H = C^* = \mathrm{Hom}_k(C,k)$ is a cocommutative Hopf k-algebra with structure maps the duals of those of C. We shall call H the Hopf dual of G.

We shall usually denote the diagonal and augmentation maps of H by Δ and ϵ, respectively, and shall on occasion find useful the so-called Heyneman-Sweedler diagonalization notation for Δ, where we write -

$$\Delta(h) = \sum_{(h)} h_{(1)} \otimes h_{(2)} \tag{2.3}$$

for h in H [21, §1.2, pp.10-12]. h is called a <u>grouplike</u> element of H if $\Delta(h) = h\otimes h$ and $\epsilon(h) = 1$. For any commutative k-algebra T, the composite map -

$$G(T) = \mathrm{Alg}_k(C,T) \hookrightarrow \mathrm{Hom}_k(C,T) \overset{\approx}{\to} H\otimes T \tag{2.4}$$

yields an injection $G(T) \hookrightarrow U(H\otimes T)$ of groups, the image of which is easily seen to be the set of all grouplike elements of the Hopf T-algebra $H\otimes T$. We shall, when convenient, identify $G(T)$ with its image in $H\otimes T$ under this natural map, thus referring to $G(T)$ as the set of grouplike elements of $H\otimes T$.

An action of a finite k-group scheme $G = \mathrm{Spec}(C)$ on a k-algebra A gives rise to a left H-module structure on A by means of the formula -

$$h(a) = \sum_{j=1}^{r} a_j\langle h, a_j'\rangle \tag{2.5}$$

for h in H and a in A, where $\langle\ \rangle: H\otimes C \to k$ is the duality pairing and $\theta(a) = \Sigma_{j=1}^{r} a_j \otimes a_j'$ in $A\otimes C$, $\theta: A \to A\otimes C$ being the k-algebra homomorphism discussed in the paragraph preceding (2.1). The k-space homomorphism $H\otimes A \to A$ corresponding to this module structure (i.e., $h\otimes a \to h(a)$) can be described more intrinsically as the image in $\mathrm{Hom}_k(H\otimes A, A)$ of θ in $\mathrm{Hom}_k(A, A\otimes C)$ under the composite of the familiar adjointness isomorphisms -

$$\mathrm{Hom}_k(A, A\otimes C) \overset{\approx}{\to} \mathrm{Hom}_k(A, \mathrm{Hom}_k(H, A)) \overset{\approx}{\to} \mathrm{Hom}_k(H\otimes A, A).$$

Such an H-module structure is called a <u>measuring</u> in [21, §7.0, pp. 137-146]; it satisfies the formulae -

(2.6) $$h(aa') = \sum_{(h)} h_{(1)}(a)h_{(2)}(a')$$

$$h(1) = \epsilon(h)\cdot 1$$

for all h in H and a, a' in A.

If, conversely, we begin with an H-module structure on A satisfying the condition (2.6), then we may reverse the above procedure to obtain a k-algebra homomorphism $\theta: A \to A\otimes C$, and this yields an action of $G = \mathrm{Spec}(C)$ on A as noted earlier.

If $G = \mathrm{Spec}(C)$ acts on A as above and T is a commutative k-algebra, then $A\otimes T$ is a left module over the T-algebra $H\otimes T$ via the H-module structure of (2.5) and base extension. $A\otimes T$ is then also a left module for the group $G(T)$ in virtue of the homomorphism $G(T) \to U(H\otimes T)$ of (2.4). It is easily verified that this operation of $G(T)$ on $A\otimes T$ coincides with that embodied in the definition of an action of G on A.

<u>Example 2.7</u>. Let $G = \mathrm{Spec}(k^{\Gamma})$ be the constant k-group scheme of (1.4), with Γ a finite group. Then the mapping $j: k\Gamma \overset{\approx}{\to} H = (k^{\Gamma})^*$ is an isomorphism of Hopf k-algebras, where

$$\langle j(\sigma),c\rangle = c(\sigma)$$

for σ in Γ, $c:\Gamma \to k$ in k^Γ, with $\langle\ ,\ \rangle: H\otimes k^\Gamma \to k$ the duality pairing. We shall identify $k\Gamma$ with H via this isomorphism. If G acts on a k-algebra A, then the operation of Γ on A arising from the left $k\Gamma$-module structure of (2.5) coincides with that arising from the action of G as in (1.5).

Assume now that $G = \mathrm{Spec}(C)$ is a finite commutative k-group scheme; hence, for each commutative k-algebra T, $G(T)$ is an abelian group. Then $H = C^*$ is a commutative k-algebra, and so $G^D = \mathrm{Spec}(H)$ is also a finite commutative k-group scheme, called the Cartier dual of G. Of course, $(G^D)^D \approx G$ in the obvious way. For example, if Γ is a finite abelian group and $G = \mathrm{Spec}(k^\Gamma)$, then $G^D = \mathrm{Spec}(k\Gamma)$ by (2.7).

An interesting generalization of quaternion algebras arises from the notion of a smash product. We shall construct it only for the case which is of interest to us here; for further information see [21, Ch.VII, §7.2, pp.153-156], [5, Ch. II, Definition 9.2, p.66], [6, §6], or [10], where its connection with a cohomological cup product pairing important in local class field theory is explored. Let $G = \mathrm{Spec}(C)$ be a finite commutative k-group scheme acting on a k-algebra A, and let $G^D = \mathrm{Spec}(H)$, with $H = C^*$, act on a k-algebra B. If $\theta_A: A \to A\otimes C$, $\theta_B: B \to B\otimes H$ are the k-algebra homomorphisms arising from this action, we shall borrow the notation of (2.3) and write -

$$\theta_A(a) = \sum_{(a)} a_{(1)}\otimes a_{(2)}$$

in $A\otimes C$ for a in A, and similarly for $\theta_B(b)$ in $B\otimes H$ with b in B. We now define a k-algebra $A\#B$ by the conditions -

(2.8a) $A\#B = A\otimes B$ as k-spaces.

Given a in A and b in B, we shall write $a\#b$ for $a\otimes b$.

(2.8b) Multiplication in $A\#B$ is defined by the formula -

$$(a\#b)(a'\#b') = \sum_{(b)} ab_{(2)}(a')\#b_{(1)}b'$$

$$= \sum_{(a'),(b)} aa'_{(1)}\langle b_{(2)}, a'_{(2)}\rangle\#b_{(1)}b'$$

where $h(a)$ is defined as in (2.5) for a in A and h in H.

The verification that $A\#B$ is indeed a k-algebra with identity element $1\#1$ is tedious but routine. The mappings $A \to A\#B$, $B \to A\#B$, where $a \to a\#1$ and $b \to 1\#b$, are injections of k-algebras, and when convenient we shall identify A, B with their images $A\#1$, $1\#B$, respectively, in $A\#B$. Hence, by slight abuse of language, we may view A and B as k-subalgebras of $A\#B$. $A\#B$ will be called the <u>smash product</u> of A and B. The k-space isomorphism $\chi = \chi_{A,B}: A\otimes B \overset{\approx}{\to} A\#B$, where $\chi(a\otimes b) = a\#b$, is not a homomorphism of k-algebras; however, it clearly has the following properties:

(2.9a) $\chi(1\otimes 1) = 1\#1$; i.e., χ preserves identity elements.

(2.9b) $\chi(uv) = \chi(u)\chi(v)$ if <u>either</u> u is in $A = A\otimes 1$ <u>or</u> v is in $B = 1\otimes B$.

For the special case in which $B = H$, the definition of the smash product $A\#H$ is meaningful even without the requirement that G be commutative, and this is the case discussed in [21, pp. 153-156], [5, Ch. II, p.66], and [6, §6]. In that case the mapping $\zeta : A\#H \to \mathrm{End}_k(A)$, where -

$$\zeta(a\#h)(a') = ah(a') \tag{2.10}$$

for a, a' in A and h in H, is a homomorphism of k-algebras, and it is not difficult to show that ζ is an isomorphism if and only if A is a PPHS for G (see, e.g., [5, Theorem 9.3, p.66]).

Now let $G = \mathrm{Spec}(C)$ be a finite commutative k-group scheme acting on a k-algebra A, and assume that $G^D = \mathrm{Spec}(H)$, with $H = C^*$, acts on a k-algebra B. If T is a commutative k-algebra, then, as noted in the paragraph preceding (2.7), $A\otimes T$ is a module over the Hopf T-algebra $H\otimes T$ in such a way that the obvious analogues of the formulae of (2.5) and (2.6) hold. Moreover, B is a right $H\otimes T$-comodule. Hence the definition of the smash product $(A\otimes T)\#_T(B\otimes T)$, as given in (2.8) is meaningful in this context, with T playing the role of k, etc. It is easy to see that the bijection $(A\#B)\otimes T \overset{\approx}{\to} (A\otimes T)\#_T(B\otimes T)$, where $(a\#b)\otimes t \to (a\otimes 1)\#_T(b\otimes t) = (a\otimes t)\#_T(b\otimes 1)$, is an isomorphism of T-algebras, and we shall identify the two algebras via this isomorphism.

In the main theorem below, we shall be concerned with the following situation.

(2.11a) $G = \mathrm{Spec}(C)$ and $\hat{G} = \mathrm{Spec}(\hat{C})$ are finite commutative k-group schemes.

(2.11b) $i = \mathrm{Spec}(j): G^D \to \hat{G}$ is a homomorphism of k-group schemes, with $j: \hat{C} \to H = C^*$ a homomorphism of Hopf k-algebras.

(2.11c) G and $\hat{G}$ act on k-algebras A and B, respectively.

Note that, under these hypotheses, G^D also acts on B via the homomorphism i, and therefore the k-algebra $A\#B$ is well-defined. This is likewise true of the k-algebra $C\#\hat{C}$, since $\hat{G}$ acts on itself by right multiplication, hence on the k-algebra $\hat{C}$, and then G^D acts on $\hat{C}$ via i.

Theorem 2.12. Let $G = \mathrm{Spec}(C)$, $\hat{G} = \mathrm{Spec}(\hat{C})$, $i = \mathrm{Spec}(j): G^D \to \hat{G}$, A, and B be as in (2.11). Then -

(a) $G \times \hat{G}$ acts on $A\#B$. If T is a commutative k-algebra, then the operation of the group $(G \times \hat{G})(T) = G(T) \times \hat{G}(T)$ on the T-algebra $(A\#B)\otimes T = (A\otimes T)\#_T(B\otimes T)$ arising from this action satisfies the formula -

$$\omega(u\#_T v) = \sigma(u)\#_T\hat{\sigma}(v)$$

for $\omega = (\sigma,\hat{\sigma})$ in $G(T) \times \hat{G}(T)$ and u in $A\otimes T$, v in $B\otimes T$.

(b) If A and B are PPHS's for G and $\hat{G}$, respectively, then $A\#B$ is a PPHS for $G \times \hat{G}$.

A proof of this theorem is given in [7, Theorem 1.20, pp.22-25].

Example 2.13. Let Γ be a cyclic group of order n with generator σ, and set $G = \mathrm{Spec}(k^\Gamma)$, in which case $G^D = \mathrm{Spec}(k\Gamma)$ as in Example 2.7 and the discussion following it. Given b in k, set $B = k[x]/(x^n-b) = k(\beta)$ (β being the image of x in B);

then B is clearly Γ-graded if we take $B_{\sigma^i} = k\beta^i$, and it is easily verified that $\mathrm{Spec}(B)$ is a PHS for G^D. If A is a normal separable field extension of k with Galois group Γ, then $\mathrm{Spec}(A)$ is a PHS for G, and $A\#B$ is then spanned as a k-space by all elements of the form $\alpha\#\beta^i$ with α in A and $0 \leq i < n$, and multiplication of such elements is given by the formula -

$$(\alpha\#\beta^i)(\alpha'\#\beta^j) = \begin{cases} \alpha\sigma^i(\alpha')\#\beta^{i+j} & \text{if } i+j < n \\ \alpha\sigma^i(\alpha')b\#\beta^k & \text{if } i+j = n+k. \end{cases}$$

It is then easy to see that $A\#B$ is the cyclic algebra arising from the cyclic field extension A of k and the element b of k as in [1, pp.82-83]. If T is a commutative k-algebra, then in view of (1.4) and (1.6) the action of $G \times G^D$ on $A\#B$ described in Theorem 2.12 corresponds to an operation of $\Gamma \times \mu_n(T)$ on $(A\#B)\otimes T$ by T-algebra automorphisms, and this is easily seen to satisfy the formula -

$$\rho\{(\alpha\#\beta^j)\otimes t\} = (\sigma^i(\alpha)\#\beta^j)\otimes\zeta^j t$$

for $\rho = (\sigma^i,\zeta)$ in $\Gamma \times \mu_n(T)$, α in A, and t in T. For the special case in which $n = 2$ and the characteristic of k is not two, there is a unique isomorphism $G \stackrel{\approx}{\to} G^D$ of k-group schemes, and so the action of $G \times G^D$ on $A\#B$ corresponds to an operation of $\Gamma \times \Gamma$ on $A\#B$ by k-algebra automorphisms. In that case $A = k(\alpha)$ with $\alpha^2 = a$ for some a in k, and we obtain the quaternion algebras of (2.2).

3. The Central Extension of a PPHS

In this section we present a correspondence between PPHS's and central extensions of group schemes. We begin with a brief description of the abelian group $\underline{Y}(G)$ which can be constructed from PPHS's for a finite commutative k-group scheme $G = \mathrm{Spec}(C)$ (see [17, Theorem 1.9, p.486] for a definition in the special case in which G is the constant k-group scheme corresponding to a finite abstract group Γ). The elements of $\underline{Y}(G)$ are the isomorphism classes of PPHS's for G, an isomorphism $A \xrightarrow{\approx} A'$ of such objects being a k-algebra isomorphism which preserves the action of G. The isomorphism class of A in $\underline{Y}(G)$ will be denoted by $\mathrm{cl}(A)$.

If A_i, for $i = 1,2$, are PPHS's for G, we set $\mathrm{cl}(A_1) + \mathrm{cl}(A_2) = \mathrm{cl}(A)$, where A is obtained as follows. Recall first that $A_1 \otimes A_2$ is a PPHS for $G \times G$ as in (2.1). We then define A to be the k-subalgebra of $A_1 \otimes A_2$ consisting of all elements a in $A_1 \otimes A_2$ satisfying the condition below -

(3.1) For each commutative k-algebra T and σ in $G(T)$, $(\sigma,\sigma^{-1})(a \otimes 1) = a \otimes 1$ in $(A_1 \otimes A_2) \otimes T$ (the element (σ,σ^{-1}) of $G(T) \times G(T) = (G \times G)(T)$ yielding a T-algebra isomorphism of $(A_1 \otimes A_2) \otimes T$ as in the discussion preceding (2.1)). $G(T)$ then operates on $A \otimes T$ by T-algebra automorphisms according to the formula $\sigma(a \otimes t) = (\sigma,1)(a \otimes t) = (1,\sigma)(a \otimes t)$, and this operation yields an action of G on A in virtue of which A can be shown to be a PPHS for G .

The operation "+" thereby obtained on $\underline{Y}(G)$ is well-defined, and renders $\underline{Y}(G)$ an abelian group. The zero element of $\underline{Y}(G)$ is cl(C), the action of G on C being given by the coalgebra structure map $C \to C \otimes C$ (i.e., by right multiplication of G on itself). The isomorphism classes of PHS's form a subgroup $\underline{X}(G)$ of $\underline{Y}(G)$ which is constructed in [5, Ch. I, §'s 3-4, pp.21-41] or [23, p.181], for example. Finally, $\underline{X}(G)$ and $\underline{Y}(G)$ are functorial in G as described in [5, Ch. I, Theorem 3.12(c)(d), pp.26-27], and the inclusion mapping $\underline{X}(G) \hookrightarrow \underline{Y}(G)$ yields a natural transformation $\underline{X} \to \underline{Y}$.

Now, if $X = \mathrm{Spec}(A)$ and $Y = \mathrm{Spec}(B)$ are PHS's for G and G^D, respectively, then A#B is a central simple k-algebra, and we obtain a homomorphism $\underline{X}(G) \otimes \underline{X}(G^D) \to \mathrm{Br}(k)$ of abelian groups, where Br(k) is the Brauer group of k and $\mathrm{cl}(A) \otimes \mathrm{cl}(B)$ is mapped to the equivalence class of A#B in Br(k). This homomorphism coincides, in a sense easily made precise, with a cohomological cup product pairing important in local class field theory. See [10] for further details on this matter.

We now consider the relation between PPHS's for G and central extensions of group schemes (for a systematic exposition of this material for PHS's see [5, Ch. III], [20], or [23]). We must first recall the (functorial) Cartier duality for finite commutative group schemes, a useful formula which describes G^D as a <u>functor</u> in terms of the functor G. If $G = \mathrm{Spec}(C)$, $H = C^*$, and T is a commutative k-algebra, we shall identify G(T) with the subgroup of $U(H \otimes T)$ consisting of all grouplike elements of the Hopf

T-algebra $H\otimes T$, as in (2.4). The group $G^D(T)$ is then similarly identified with the corresponding subgroup of $U(C\otimes T)$, since $H^* = C^{**} = C$. The duality pairing $\langle\, ,\rangle : H\otimes C \to k$ yields, by base extension, a pairing $\langle\, ,\rangle_T : (H\otimes T)\otimes_T (C\otimes T) \to T$, and hence a mapping $\langle\, ,\rangle_T : G(T)\times G^D(T) \to T$. Recall finally that, since G acts on itself by right multiplication, G acts on the k-algebra C, and hence $C\otimes T$ is a left $H\otimes T$-module as described in the paragraph preceding (2.7). In particular, if σ,φ are in $G(T)\subseteq H\otimes T$ and $G^D(T)\subseteq C\otimes T$, respectively, then $\sigma(\varphi)$ is a well-defined element of $C\otimes T$.

<u>Proposition 3.2</u>. Let $G = \mathrm{Spec}(C)$ be a finite commutative k-group scheme, $H = C^*$, and let G_m denote the multiplicative k-group scheme; i.e., $G_m(T) = U(T)$ for T a commutative k-algebra. We then have an isomorphism -

$$G^D(T) \xrightarrow{\approx} \mathrm{Hom}(G^T, G_m^T)$$

of abelian groups which is natural in T and G, where G^T, G_m^T denote the restrictions of the functors G, G_m, respectively, to the category of commutative T-algebras, and $\mathrm{Hom}(G^T, G_m^T)$ is the abelian group of functor homomorphisms (i.e., additive natural transformations) from G^T to G_m^T. This isomorphism has the following properties -

(a) Let $\bar\varphi : G^T \to G_m^T$ denote the image in $\mathrm{Hom}(G^T, G_m^T)$ of φ in $G^D(T)$. If S is a commutative T-algebra, then $\bar\varphi_S : G(S) = G^T(S) \to G_m^T(S) = G_m(S)$ maps σ in $G(S)$ to $\langle\sigma',\varphi\rangle_S$ in $G_m(S)\subseteq S$, with $\langle\ ,\ \rangle_S : G(S)\times G^D(S)\to S$ as in the preceding paragraph, and σ' the image of σ in $G(S)$ under the natural map $G(T)\to G(S)$.

(b) Given σ in $G(T)\subseteq H\otimes T$ and φ in $G^D(T)\subseteq C\otimes T$, then $\sigma(\varphi)=\langle\sigma,\varphi\rangle_T\varphi$ in $C\otimes T$.

The isomorphism $G^D(T)\xrightarrow{\approx}\mathrm{Hom}(G^T,G_m^T)$ is well-known (see e.g., [8, pp.156-159] or [18, §3, pp.423-424]), and the explicit description of it given in (a) follows easily from the construction given in these references. (b) is then a consequence of (a) via a routine computation using (2.5) and the fact that φ is a group-like element of $C\otimes T$.

Remarks 3.3. Of course, the functor G^T of Proposition 3.2 is the T-group scheme $\mathrm{Spec}(C\otimes T)$, and a similar statement holds for G_m. We shall, when useful, identify $G^D(T)$ with $\mathrm{Hom}(G^T,G_m^T)$ via the natural isomorphism of Proposition 3.2 and shall, by convenient abuse of notation, denote the image in $G(S)$ of an element σ in $G(T)$ by the same letter σ, for S a commutative T-algebra.

Now let $G=\mathrm{Spec}(C)$ be as above, and A be a PPHS for G. We shall construct a central extension -

$$(3.4)\qquad \xi_A : 1\to G_m\to E_A\to G^D\to 1$$

of the functor G^D by the functor G_m (i.e., ξ_A can be viewed as an extension of presheaves on the category of affine k-schemes; over an arbitrary ground ring k one must go to extensions of sheaves in a sort of Zariski topology, or any finer topology [5, Ch. III, §15, pp.96-102][23]).

Let, for the moment, T be a commutative k-algebra, and a be an invertible element of $A \otimes T$. If S is a commutative T-algebra and σ is in $G(S)$, then we define an invertible element $\varphi_S^a(\sigma)$ of $A \otimes S$ by the formula -

$$\varphi_S^a(\sigma) = a^{-1}\sigma(a) \tag{3.5}$$

where, when we write "a" on the right hand side of this formula, we really mean the image of a in $A \otimes S$ under the natural map $A \otimes T \to A \otimes S$. We then define a subset $E_A(T) \subseteq U(A \otimes T)$ by the condition -

$$E_A(T) = \{a \text{ in } U(A \otimes T) \mid \varphi_S^a(\sigma) \text{ is in } G_m(S) \subset U(A \otimes S) \tag{3.6}$$

for every commutative T-algebra S and σ in $G(S)\}$.

It is easily verified that $E_A(T)$ is a subgroup of $U(A \otimes T)$ and $G_m(T) \subseteq E_A(T) \subseteq U(A \otimes T)$. Since $A \otimes T$ is a T-algebra, $G_m(T)$ is a central subgroup of $E_A(T)$. In addition, if a is in $E_A(T)$, then (3.5) yields a homomorphism $\varphi_S^a : G(S) \to G_m(S)$ which is natural in S, and hence we obtain an element $\varphi^a : G^T \to G_m^T$ in $\mathrm{Hom}(G^T, G_m^T) = G^D(T)$. The mapping $a \to \varphi^a$ is then a homomorphism $E_A(T) \to G^D(T)$ of groups which is natural in T, and hence provides a homomorphism $E_A \to G^D$.

<u>Theorem 3.7</u>. Let $G = \mathrm{Spec}(C)$ be a finite commutative k-group scheme.

(a) If A is a PPHS for G, then the sequence ξ_A of (3.4), with the homomorphism $E_A \to G^D$ defined as above, is an exact sequence of group-valued functors on the category of commutative k-algebras, and yields a central extension of G^D by G_m. Moreover, E_A is representable, and hence is a k-group scheme.

(b) Let $\mathrm{Cext}(G^D, G_m)$ denote the group of isomorphism classes of central extensions of G^D by G_m, and $\mathrm{Ext}(G^D, G_m)$ denote the subgroup of isomorphism classes of abelian extensions. Then the mapping $\mathrm{cl}(A) \to \mathrm{cl}(\xi_A)$ yields an isomorphism $\underline{Y}(G) \xrightarrow{\approx} \mathrm{Cext}(G^D, G_m)$ of abelian groups, where in each case "cl" means "isomorphism class of". This isomorphism maps $\underline{X}(G) \subset \underline{Y}(G)$ onto $\mathrm{Ext}(G^D, G_m)$.

A proof of the theorem for PHS's and abelian extensions can be found in [5, Ch. III, Theorem 16.14, p.109]; this proof carries through in the context outlined here.

Remarks 3.8. (a) For alternate proofs in the abelian case see [20] and [23]. Unfortunately, we know of no counterpart, in the present context, of the elegant geometric arguments of [23]. For the relation of the theorem to classical Kummer theory see [5, Ch. III, §13, pp.84-87].

(b) Cohomological versions of Theorem 3.7, for the special case in which G is a constant k-group scheme, appear in [11] and [4, Theorem 5.8, p.684].

In the next definition we forget (for the moment) all about group schemes, algebras, etc., and deal with ordinary abelian groups.

Definition and Remarks 3.9. Let A, B, C be abelian groups, and $\langle\, , \rangle : A \otimes_{\mathbb{Z}} B \to C$ be a pairing (i.e., a homomorphism where we denote the image of $a \otimes b$ in C by $\langle a,b \rangle$). We define a (usually nonabelian) group E as follows. As a set, $E = C \times A \times B$, with group law satisfying the formula -

$$(c,a,b)(c',a',b') = (c+c'+\langle a',b \rangle, a+a', b+b').$$

It is easily verified that E is a group with identity element $1 = (0,0,0)$, and the sequence -

$$1 \to C \to E \to A \times B \to 1$$

yields a central extension of $A \times B$ by C, where $c \to (c,0,0)$ for c in C and $(c,a,b) \to (a,b)$ for a in A, b in B. This will be called the Heisenberg extension of $A \times B$ by C (relative to the given pairing).

Examples and Variations 3.10. (a) Let $A = B = C = \mathbb{Z}_2$, and the pairing $\langle , \rangle : A \otimes_{\mathbb{Z}} B \to C$ be the unique non-degenerate one. Then E is the dihedral group of order 8 .

(b) It is clear that Heisenberg extensions can be defined for abelian group objects in any category in which the notion of pairing can be made meaningful. For example, they can be constructed for topological groups, algebraic groups, and Γ-modules, with Γ an abstract group.

(c) The central extensions of (3.9), in the context of locally compact abelian groups, play an important role in Weil [24, Ch. I, §4, p.149]. The classical Heisenberg (algebraic) group is obtained by taking $A = V$, $B = V^*$, $C = k$, with V a k-space of finite dimension and $\langle\ ,\ \rangle : V \otimes V^* \to k$ the duality pairing.

(d) We shall be primarily interested in the Heisenberg extensions for finite commutative k-group schemes, the construction of which is rendered possible by the Cartier duality formula of Proposition 3.2. Namely, let $G = \mathrm{Spec}(C)$, $\hat{G} = \mathrm{Spec}(\hat{C})$ be finite commutative k-group schemes, and $\rho : \hat{G} \to G^D$ be a homomorphism of k-group schemes. Given a commutative k-algebra T, we have the homomorphism -

$$\rho_T : \hat{G}(T) \to G^D(T) = \mathrm{Hom}(G^T, G_m^T)$$

of abelian groups, where we use Proposition 3.2 to identity $G^D(T)$ with $\mathrm{Hom}(G^T, G_m^T)$. Given σ in $G(T)$ and $\hat{\sigma}$ in $\hat{G}(T)$, we shall write -

$$(\sigma, \hat{\sigma})_T = \langle \sigma, \rho_T(\hat{\sigma}) \rangle_T$$

in $G_m(T)$, the right-hand side being as in the paragraph preceding Proposition 3.2. $(\ ,\ )_T : G(T) \times \hat{G}(T) \to G_m(T)$ is then a (bimultiplicative) pairing and is natural in T. The corresponding Heisenberg extensions -

$$1 \to G_m(T) \to E(T) \to G(T) \times \hat{G}(T) \to 1$$

then define a central extension of functors E of $G \times \hat{G}$ by G_m, which we call the Heisenberg extension of $G \times \hat{G}$ by G_m relative to the homomorphism $\rho : \hat{G} \to G^D$. By a standard argument, E is representable and hence a k-group scheme.

The main theorem below relates the two concepts of smash product and Heisenberg extension.

Theorem 3.11. Let $G = \mathrm{Spec}(C)$, $\hat{G} = \mathrm{Spec}(C)$ be finite commutative k-group schemes, and $i: G^D \to G$ be a k-group scheme homomorphism. Let A, B be PPHS's for G and $\hat{G}$, respectively, and ξ_A, ξ_B be the corresponding central extensions as in (3.4). View $A\#B$ as a PPHS for $G \times \hat{G}$ as in Theorem 2.12. Then the isomorphism -

$$\underline{Y}(G \times \hat{G}) \xrightarrow{\approx} \mathrm{Cext}((G \times \hat{G})^D, G_m)$$

of Theorem 3.7 maps $\mathrm{cl}(A\#B)$ to $\mathrm{cl}(\bar{\xi}_A) + \mathrm{cl}(\bar{\xi}_B) + \mathrm{cl}(\Xi)$, where -

(a) $\mathrm{cl}(\bar{\xi}_A)$ is the image in $\mathrm{Ext}((G \times \hat{G})^D, G_m)$ of $\mathrm{cl}(\xi_A)$ under the homomorphism $\mathrm{Cext}(G^D, G_m) \to \mathrm{Cext}((G \times \hat{G})^D, G_m)$ induced by the projection $(G \times \hat{G})^D = G^D \times \hat{G}^D \to G^D$, and $\mathrm{cl}(\bar{\xi}_B)$ is similarly defined.

(b) $\Xi : 1 \to G_m \to E \to G^D \times \hat{G}^D = (G \times \hat{G})^D \to 1$ is the Heisenberg extension of $G^D \times \hat{G}^D$ by G_m relative to the homomorphism $i^D : \hat{G}^D \to (G^D)^D$.

In particular, the above isomorphism maps $\mathrm{cl}(C\#\hat{C})$ in $\underline{Y}(G \times \hat{G})$ to $\mathrm{cl}(\Xi)$ in $\mathrm{Cext}((G \times \hat{G})^D, G_m)$.

See [7, Theorem 6.12, pp.79-86] for a proof of this result. Recently M. Takeuchi [22] has obtained generalizations of Theorem 3.7 to arbitrary affine and formal k-groups, and of Theorem 3.11 to affine and formal commutative k-groups.

4. The Witt Group of Regular PHS's.

In this section we shall exploit the analogy, elucidated in [7], between PPHS's and bilinear forms in order to define and investigate a counterpart, in our context, of the Witt group of such forms (see, e.g., [14, Ch. Two, §1, pp. 34-37] or [13]). Throughout our discussion, k will be a field, and PPHS(k) will denote the category of which:

(4.1a) The objects are pairs (G, A), with G a finite commutative k-group scheme and A a PPHS for G.

(4.1b) The morphisms are pairs (i,φ): $(G,A) \to (G',A')$, where $i: G \to G'$ is an isomorphism of k-group schemes and $\varphi: A \xrightarrow{\approx} A'$ is an isomorphism of k-algebras which is also an isomorphism of PPHS's for G if A' is viewed as such with G-action induced by i.

Note that the morphisms in the category PPHS(k) are all isomorphisms. Moreover, PPHS(k) is a category with product $\otimes$ in the sense of [2, Ch. VII, §1, pp. 344-353], where we use (2.1) to define-

$$(4.2) \qquad (G_1, A_1) \otimes (G_2, A_2) = (G_1 \times G_2, A_1 \otimes A_2).$$

We now single out for special attention certain classes of PPHS's which we shall view as analogues, in our context, of the non-degenerate and metabolic bilinear forms, respectively [13, p. 122].

Definition and Remarks 4.3. (a) An object (G, A) of PPHS(k) will be called regular if there exist finite commutative k-group schemes G' and G_1, PPHS's. A' and A_1 for G' and G_1, respectively, and a commutative PPHS B_1 for G_1^D, such that-

$$(G,A) \otimes (G',A') \approx (G_1 \times G_1^D,\ A_1 \# B_1)$$

in PPHS(k) (of course, to say that B_1 is commutative is simply to say that $Y_1 = \mathrm{Spec}(B_1)$ is an ordinary PHS for G_1^D).

(b) An object of PPHS(k) will be called metabolic if it is isomorphic to one of the form $(G \times G^D,\ A\#H)$, with $G = \mathrm{Spec}(C)$, $H = C^*$, and A a PPHS for G.

Terminology 4.4. If (G,A) is a regular (metabolic) object of PPHS(k), we shall say that A is a regular (metabolic) PHS for G. Note that, in dropping the essentially superfluous prescript "pseudo-", we are abusing language somewhat in that a regular or metabolic PHS A is never actually a PHS for G unless G is trivial, since A is a commutative algebra only in that case.

The proposition below is useful in constructing regular and metabolic PHS's.

Proposition 4.5. (a) Let G_n, $\hat{G}_n$ be finite commutative k-group schemes for $n = 1,2$, and $j_n : G_n^D \to \hat{G}_n$ be a k-group scheme homomorphism. Let A_n, B_n be PPHS's for G_n and $\hat{G}_n$, respectively. Then

$$(i,\omega) : (G_1 \times G_2 \times \hat{G}_1 \times \hat{G}_2,\ (A_1 \times A_2)\#(B_1 \otimes B_2)) \xrightarrow{\approx}$$

$$(G_1 \times \hat{G}_1 \times G_2 \times \hat{G}_2,\ (A_1 \# B_1) \otimes (A_2 \# B_2))$$

is an isomorphism in the category PPHS(k), where-

$$i: G_1 \times G_2 \times \widehat{G}_1 \times \widehat{G}_2 \xrightarrow{\approx} G_1 \times \widehat{G}_1 \times G_2 \times \widehat{G}_2$$

is obtained by interchanging the two middle factors, and-

$$\varphi:(A_1 \otimes A_2)\#(B_1 \otimes B_2) \xrightarrow{\approx} (A_1\#B_1) \otimes (A_2\#B_2)$$

satisfies the formula-

$$\varphi\{(a_1 \otimes a_2)\#(b_1 \otimes b_2)\} = (a_1\#b_1) \otimes (a_2\#b_2).$$

Each of the assertions below follows, in a routine fashion, either from the basic definitions or from the proposition just stated.

Corollary 4.6. (a) A metabolic PHS is regular.

(b) If A_i is a regular PHS for G_i $(i = 1,2)$, then $A_1 \otimes A_2$ is a regular PHS for $G_1 \times G_2$.

(c) If A and B are any PPHS's for G and G^D, respectively, and B is commutative, then $A\#B$ is a regular PHS for $G \times G^D$.

In view of Corollary 4.6(a), the full subcategories Reg(k) and Met(k) of PPHS(k), the objects of which are, respectively, the regular and metabolic objects of PPHS(k), are themselves categories with product; namely, the same product as in PPHS(k). Moreover, the inclusion functor μ:Met(k) → Reg(k) is a functor of such categories. Hence, passing to the Grothendieck groups of these categories [2, p. 346] we obtain a homomorphism-

$$K(\mu): K(Met(k)) \to K(Reg(k))$$

of abelian groups.

Definition 4.7. We define $Z(k)$, the Witt group of regular PHS's over k, to be the cokernel of $K(\mu)$. Hence we have the exact sequence of abelian groups-

$$K(Met(k)) \xrightarrow{K(\mu)} K(Reg(k)) \to Z(k) \to 0.$$

The final result of this section concerns the relation between $Z(k)$ and the Brauer group $Br(k)$ of equivalence classes of Azumaya k-algebras. If (G, A) is an object of $Reg(k)$, we denote the corresponding element of $Z(k)$ by $cl(G, A)$. Similarly, if B is an Azumaya k-algebra, we write its equivalence class in $Br(k)$ as $cl(B)$.

Theorem 4.8. There exists a homomorphism

$$\alpha_k : Z(k) \to Br(k)$$

such that $\alpha_k(cl(G, A)) = cl(A)$. The image of α_k contains (at least) all elements $cl(B)$ of $Br(k)$ with B a cyclic k-algebra. In particular, α_k is surjective if k is a local or global field.

The existence of α_k follows from an easy generalization of the arguments of [10], together with the fact that the mapping of (2.10) is an isomorphism for PPHS's. The assertion of the theorem regarding the image of α_k is an immediate consequence of (2.13). We do not know whether α_k is always surjective.

The abelian group Z(k) discussed here bears a superficial resemblance to other variants of the Brauer group defined using group or Hopf algebra actions, such as the equivariant Brauer group of Fröhlich-Wall [9] (see also [15]), and the Brauer group of H-dimodule algebras developed by F.W. Long [16], with H a commutative, cocommutative Hopf k-algebra. There are, however, distinct and important differences between these and Z(k). For example, the Brauer group of H-dimodule algebras contains, as a subgroup, the ordinary Brauer group Br(k), the embedding being obtained by viewing an Azumaya k-algebra as an H-dimodule algebra with trivial H-action [16, p. 588-589]. There appears to be no corresponding map Br(k) → Z(k), since there is no reason to expect that an arbitrary Azumaya k-algebra could be made into a PPHS for some finite commutative k-group scheme, and even if it could, the group scheme would not be uniquely determined. Moreover, there is an H-dimodule-algebra Brauer group for each choice of H, whereas the definition of Z(k) involves regular PHS's for all finite commutative k-group schemes simultaneously.

5. Regular PHS's and Symplectic Pairings

We shall now describe a homomorphism from $Z(k)$ to a Witt group constructed from symplectic pairings of finite commutative k-group schemes. Some of the ideas of this section were anticipated by Hoechsmann [12].

By a pairing on a finite commutative k-group scheme G we mean a k-scheme morphism $\beta: G \times G \to G_m$ which is bimultiplicative in the obvious sense. Hence, for each commutative k-algebra T, the mapping $\beta_T: G(T) \times G(T) \to G_m(T)$ satisfies the axioms -

$$(5.1a) \qquad \beta_T(\sigma_1\sigma_2,\tau) = \beta_T(\sigma_1,\tau)\beta_T(\sigma_2,\tau)$$
$$\beta_T(\sigma,\tau_1\tau_2) = \beta_T(\sigma,\tau_1)\beta_T(\sigma,\tau_2)$$

for all σ, σ_i, τ, τ_i in $G(T)$ $(i = 1,2)$. β will be called symplectic if, for each T as above -

$$(5.1b) \qquad \beta_T(\sigma,\sigma) = 1$$

for all σ in $G(T)$. β will be called nondegenerate if, for all T, the mappings $G(T) \to G^D(T) = \mathrm{Hom}(G^T, G_m^T)$ induced by β; namely

$$(5.1c) \qquad \sigma \to \beta_T(\sigma,-),\ \tau \to \beta_T(-,\tau) \quad (\sigma,\tau \text{ in } G(T))$$

are bijective. If β_i is a pairing on G_i $(i = 1,2)$, then the orthogonal product of β_1 and β_2 is the pairing on $G = G_1 \times G_2$ defined by the formula -

(5.1d) $\quad \beta_T\{(\sigma_1,\sigma_2),(\tau_1,\tau_2)\} = (\beta_1)_T(\sigma_1,\tau_1)(\beta_2)_T(\sigma_2,\tau_2).$

Finally, a symplectic pairing γ on $G \times G^D$ will be called metabolic if there exists a symplectic pairing β on G such that -

(5.1e) $\quad \gamma_T\{(\sigma,\tau),(\sigma',\tau')\} = \beta_T(\sigma,\sigma')(\sigma,\tau')_T(\sigma',\tau)_T^{-1}$

for each T and σ, σ' in $G(T)$ and τ, τ' in $G^D(T)$, with $(\ ,\ ): G \times G^D \to G_m$ the Cartier duality pairing of Proposition 3.2.

Lemma 5.2. A metabolic symplectic pairing is non-degenerate.

Now consider the category $Sp(k)$, of which -

(5.3a) An object is a pair (G,β), with G a finite commutative k-group scheme and β a symplectic non-degenerate pairing on G;

(5.3b) A morphism $i:(G,\beta) \to (G',\beta')$ is an isomorphism $i: G \xrightarrow{\approx} G'$ of k-group schemes such that the diagram below commutes

$$\begin{array}{ccc} G \times G & \xrightarrow{i \times i} & G' \times G' \\ & \searrow^{\beta} \quad \swarrow_{\beta'} & \\ & G_m & \end{array}$$

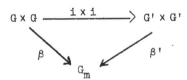

$Sp(k)$ is a category with product [2, p.344]; namely, the orthogonal product of (5.1d). Moreover, in view of Lemma 5.2, the metabolic symplectic pairings form a full subcategory Metsp(k) which is easily seen to be closed under this product. Hence, passing to the Grothendieck groups of these categories, we obtain a homomorphism -

$$K(\nu): K(\mathrm{Metsp}(k)) \to K(\mathrm{Sp}(k))$$

of abelian groups, with $\nu: \mathrm{Metsp}(k) \hookrightarrow \mathrm{Sp}(k)$ the inclusion functor.

Definition 5.4. We define $W_{sp}(k)$, the Witt group of non-degenerate symplectic pairings over k, to be the cokernel of $K(\nu)$. Hence we have an exact sequence of abelian groups -

$$K(\mathrm{Metsp}(k)) \xrightarrow{K(\nu)} K(\mathrm{Sp}(k)) \to W_{sp}(k) \to 0 .$$

We are now ready to describe a connection between symplectic pairings and regular PHS's. Let A be a regular PHS for a finite commutative k-group scheme G, and let -

$$\xi_A: 1 \to G_m \to E \to G^D \to 1$$

be its corresponding central extension as in (3.4). The commutator operation on E then yeidls a k-scheme morphism $\beta^A: G^D \times G^D \to G_m$. That is, if T is a commutative k-algebra and σ, τ are in $G^D(T)$, then -

$$\beta^A_T(\sigma,\tau) = \bar{\sigma}\bar{\tau}\bar{\sigma}^{-1}\bar{\tau}^{-1} \tag{5.5}$$

with $\bar{\sigma}, \bar{\tau}$ in $E(T)$ mapping onto σ and τ, respectively. It is easily verified that β^A is a symplectic pairing on G^D.

<u>Theorem 5.6</u>. There exist abelian group homomorphisms $K(\mathrm{Reg}(k)) \to K(\mathrm{Sp}(k))$, $K(\mathrm{Met}(k)) \to K(\mathrm{Metsp}(k))$, and $\omega_k: Z(k) \to W_{sp}(k)$ such that the diagram below commutes -

$$\begin{array}{ccccccc} K(\mathrm{Met}(k)) & \xrightarrow{K(\mu)} & K(\mathrm{Reg}(k)) & \longrightarrow & Z(k) & \longrightarrow & 0 \\ \downarrow & & \downarrow & & \downarrow{\scriptstyle \omega_k} & & \\ K(\mathrm{Metsp}(k)) & \xrightarrow{K(\nu)} & K(\mathrm{Sp}(k)) & \longrightarrow & W_{sp}(k) & \longrightarrow & 0 \end{array}$$

and the upper and lower exact sequences as in (4.7) and (5.4), respectively. Given $\mathrm{cl}(G,A)$ in $K(\mathrm{Reg}(k))$ or $K(\mathrm{Met}(k))$, its image in $K(\mathrm{Sp}(k))$ or $K(\mathrm{Metsp}(k))$ is $\mathrm{cl}(G^D,\beta^A)$.

Theorem 5.6 follows from Theorem 3.11 and computations of a fairly routine nature. It can be used to identify $Z(k)$ in the easiest special case -

<u>Theorem 5.7</u>. $Z(k) = 0$ if k is an algebraically closed field of characteristic zero.

References

1. E. Artin, C. Nesbitt, R. Thrall, Rings with minimum condition, University of Michigan Press, Ann Arbor 1944.

2. H. Bass, Algebraic K-theory, W.A. Benjamin, Inc., New York 1968.

3. ______, Unitary algebraic K-theory, In: Algebraic K-Theory III (Hermitian K-Theory and Geometric Applications), Lecture Notes in Math. no. 343, Springer-Verlag, Berlin-Heidelberg-New York 1973.

4. S. Chase and A. Rosenberg, A theorem of Harrison, Kummer theory, and Galois algebras, Nagoya Math. J. vol. 27 (1966), pp.663-685.

5. S. Chase and M. Sweedler, Hopf algebras and Galois theory, Lecture Notes in Math. no. 97, Springer-Verlag, Berlin-Heidelberg-New York 1969.

6. S. Chase, Infinitesimal group scheme actions on finite field extensions, to appear in Amer. J. Math.

7. ______, On principal homogeneous spaces and bilinear forms, preprint (1975).

8. M. Demazure and P. Gabriel, Groupes algebrique (Tome I), North Holland Publishing Co., Amsterdam 1970.

9. A. Fröhlich and C.T.C. Wall, Equivariant Brauer groups in algebraic number theory, Bull. Soc. Math. France 25 (1971), pp.91-96.

10. J. Gamst and K. Hoechsmann, Quaternions generalises, Compt. Rendu. Acad. Sci. Paris 269(1969), pp.560-562.

11. D. Harrison, Abelian extensions of arbitrary fields, Trans. Amer. Math. Soc. 52 (1963), pp.230-235.

12. K. Hoechsmann, Über nicht-kommutative abelsche algebren, J. Reine Angw. Math. 218 (1965), pp.1-5.

13. M. Knebusch, A. Rosenberg, and R. Ware, Structure of Witt rings and quotients of abelian group rings, Amer. J. Math. 54 (1972), pp.119-155.

14. T.-Y. Lam, The algebraic theory of quadratic forms, W.A. Benjamin, Inc., New York 1973.

15. F. Long, A generalization of the Brauer group of graded algebras, Proc. London Math. Soc. 29 (1974), pp.237-256.

16. ______, The Brauer group of dimodule algebras, J. of Algebra 30 (1974), pp.559-601.

17. M. Orzech, A cohomological description of abelian Galois extensions, Trans. Amer. Math. Soc. 137 (1969), pp.481-499.

18. S. Shatz, Cohomology of Artinian group schemes over local fields, Ann. Math. 79 (1964), pp.411-449.

19. ______, Galois theory, In: Category theory, homology theory and their applications (Vol. I), Lecture Notes in Math. no. 86, Springer-Verlag, Berlin-Heidelberg-New York 1969.

20. ______, Principal homogeneous spaces for finite group schemes, Proc. Amer. Math. Soc. 22 (1969), pp.678-680.

21. M. Sweedler, Hopf algebras, W.A. Benjamin, Inc., New York 1969.

22. M. Takeuchi, private communication.

23. W. Waterhouse, Principal homogeneous spaces and group scheme extensions, Trans. Amer. Math. Soc. 153 (1971), pp.181-189.

24. A. Weil, Sur certains groupes d'opérateurs unitaires, Acta Math. 111 (1964), pp.143-211.